MAGIC WITH MATH

Exploring Number Relationships and Patterns

Written by Paul Swan

Published by Didax Educational Resources
www.didax.com

Published with the permission of R.I.C. Publications Pty. Ltd.

Copyright © 2003 by Didax, Inc., Rowley, MA 01969. All rights reserved.

First published by R.I.C. Publications Pty. Ltd., Perth, Western Australia. Revised by Didax Educational Resources.

Limited reproduction permission: The publisher grants permission to individual teachers who have purchased this book to reproduce the blackline masters as needed for use with their own students. Reproduction for an entire school or school district or for commercial use is prohibited.

Printed in the United States of America.

Order Number 2-162
ISBN 978-1-58324-155-4

B C D E F 11 10 09 08 07

395 Main Street
Rowley, MA 01969
www.didax.com

Foreword

Magic with Math has been designed to enrich the mathematical experience of older students or simply enthral younger students with the seemingly magical properties of mathematics. Teachers may use the ideas to introduce lessons, to finish a lesson on "a high note," to arouse curiosity, or extend students' thinking. Alternatively, teachers may wish to copy the activities and allow students to jot down ideas or the calculations required to complete the trick. Older students should be encouraged to look at the reason behind the tricks, asking, what makes them work? In many cases, this will provide the opportunity to discuss and develop algebraic ideas and concepts. Younger students who are not ready for algebra may simply be allowed to wonder at the magic of mathematics. Remember, the purpose of introducing a "mathemagical" trick should not be to baffle students with the intricacies of algebra but to turn them on to mathematics and motivate them!

Contents

Teacher's Notes	4
Activity Overview	5–6
Student Self-assessment Outline	7
Student Checklist	8
Magic Numbers – Introduction	9
Magic Numbers – Activities	10–16
Magic Numbers – Explanations and Answers	17–18
Think of a Number – Introduction	20
Think of a Number – Activities	21–24
Think of a Number – Explanations and Answers	25
Properties of Nine – Introduction	27
Properties of Nine – Activities	28–34
Properties of Nine – Explanations and Answers	35–36
Algebra-based Tricks – Introduction	37
Algebra-based Tricks – Activities	39–48
Algebra-based Tricks – Explanations and Answers	49–51
Card Tricks – Introduction	53
Card Tricks – Activities	54–56
Card Tricks – Explanations and Answers	57
A Mixed Bag – Introduction	59
A Mixed Bag – Activities	60–64
A Mixed Bag – Explanations and Answers	65–66
Teacher Tricks – Introduction	68
Teacher Tricks – Activities and Explanations	69–86

Teacher's Notes

The activities in **Magic with Math** can be used in a variety of ways. Teachers may use the ideas to introduce lessons, to finish a lesson on a "high note," to arouse curiosity, or to extend students' thinking.

The tricks have been grouped under broad headings to assist teachers wishing to use the activities as part of a lesson or sequence of lessons. Many of the activities rely on number patterns and algebra; however, a knowledge of algebra is not required to experience the "mathemagic" in the tricks. It will be helpful to introduce some basic algebraic principles to the students before they attempt to describe how the tricks work and then create their own number tricks.

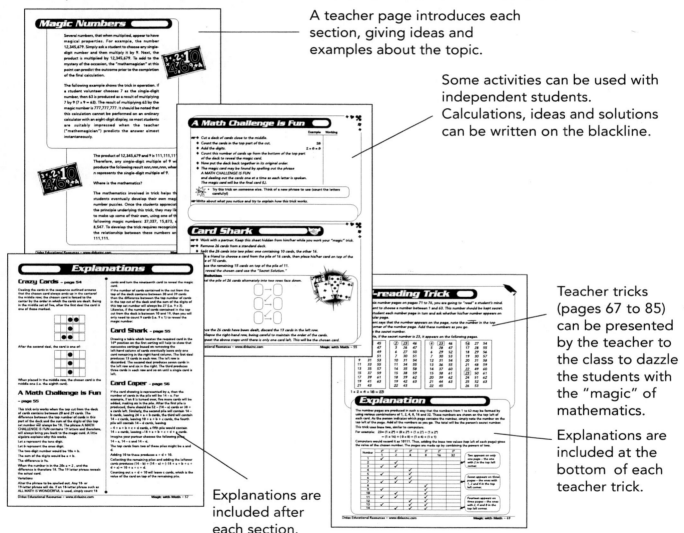

A teacher page introduces each section, giving ideas and examples about the topic.

Some activities can be used with independent students. Calculations, ideas and solutions can be written on the blackline.

Teacher tricks (pages 67 to 85) can be presented by the teacher to the class to dazzle the students with the "magic" of mathematics.

Explanations are included at the bottom of each teacher trick.

Explanations are included after each section.

Materials needed to complete activities in **Magic with Math** include:
- dice (six-sided)
- decks of cards
- copy paper
- matchboxes
- rulers
- dominoes
- calculators
- scissors
- matchboxes
- pieces of string (approx. 1 yard long)

A calendar (page 39) and 8x8 grids (page 60) have been provided for use with the activities where required.

Activity Overview

The following table outlines the activities in **Magic with Math**.

Page	Title	Who	Area(s) of Mathematics
9	Magic Numbers Section	teacher	
10	Thirty-seven	student	number patterns
10	Number Novelties	student	algebra
11	1,089	student	pattern and algebra
11	1,089 Revisited	student	pattern and algebra
12	Tautonyms	student	number patterns
12	Triple Treat	student	number patterns
13	Criss Cross I	student	number patterns
13	Criss Cross II	student	number patterns
14	Crazy Calculator	student	number patterns
14	Double up	student	number patterns
15	The Missing Eight	student	number patterns
15	Four-digit Madness	student	number patterns
16	Palindromes	student	number patterns
16	Multiple Madness	student pairs	number patterns
20	Think of a Number Section	teacher	
21	Think of a Number	student	simple algebra
21	Give Me Five	student	simple algebra
22	Diabolical Dominoes	student pairs	simple algebra
22	Dice Detective	student pairs	simple algebra
23	Pick a Card	student pairs	simple algebra
23	Deck Detective	student pairs	simple algebra
24	Create Your Own	student	simple algebra
27	Properties of 9 Section	teacher	
28	Multiples of Nine	student	multiples of nine
28	Who Knows?	student	properties of nine, algebra
29	Mind-reader I	student pairs	properties of nine, algebra
29	Mind-reader II	student pairs	properties of nine, algebra
30	Number ESP	student pairs	properties of nine
31	Missing Number	student pairs	properties of nine
31	Digit Detection	student pairs	properties of nine
32	Scrambled Digits I	student	properties of nine, algebra
32	Scrambled Digits II	student pairs	properties of nine
33	Double-digit Dilemma	student	properties of nine, algebra
33	Ten Times	student	properties of nine, algebra
34	Role Reversal	student	properties of nine, algebra
34	Mystery Matchbox	student pairs	properties of nine
38	Algebra-based Tricks	teacher	
40	Calculation Shortcuts I	student	algebra, differences of squares
40	Calculation Shortcuts II	student	algebra
41	Square	student	algebra
41	A Trick for Squares	student	algebra
42	Digit Shuffle	student	algebra

Activity Overview

Page	Title	Who	Area(s) of Mathematics
42	Finding the Lost Digit	student	algebra
43	Triple-digit Division	student	algebra
43	Seeing Double	student	algebra
44	Do or Die	student	algebra, properties of fair dice
44	Dice Dropping	student	algebra
45	Double Trouble	student	algebra
45	3,087	student	algebra
46	Cycling Digits	student	algebra
46	Mathemagic	student	algebra
47	Calendar Magic	student	algebra, uniform calendar layout
47	Magic Months	student pairs	algebra, uniform calendar layout
48	Cool Calendar	student	algebra, uniform calendar layout
48	Your Puzzle	student	algebra, uniform calendar layout
53	Card Tricks	teacher	
54	Crazy Cards	student pairs	sequencing
55	A Math Challenge is Fun	student	algebra
55	Card Shark	student pairs	sequencing
56	Card Caper	student pairs	algebra
59	Mixed Bag	teacher	
61	Stepping Through Paper	student	topology
61	Tied up in Knots	student pairs	topology
62	A Shape with a Twist I	student	topology
62	A Shape with a Twist II	student	topology
63	Missing Area I	student	pattern
63	Number Words	student pairs	properties of fair dice
64	Missing Area II	student	area of a rectangle
64	Thrice Dice	student pairs	area triangle/trapezoid
68	Teacher Tricks	teacher	
69 – 76	Mind-reading Trick	teacher	powers of two
77 – 79	Number Strips	teacher	addition
80 – 81	Tricky Tables	teacher	addition, subtraction, multiplication
82	Lightning Addition	teacher	Fibonnaci sequence, addition, pattern
83	Super Sequences	teacher	sequence, pattern, addition, multiplication
84	Mystifying Multiplication	teacher	addition
85	Number Spelling	teacher	pattern, sequence
86	It all Adds up	teacher	subtraction, multiplication

Self-assessment Outline

Name: _____ Date: / /

Math Activity: _____

What I did:

To help me, I used the following materials:

I learned the following things:

One thing I found difficult was: _____

The activity made me feel: _____

Self-assessment Outline

Name: _____ Date: / /

Math Activity: _____

What I did:

To help me, I used the following materials:

I learned the following things:

One thing I found difficult was: _____

The activity made me feel: _____

Student Checklist

Activity	Completed ✓	Activity	Completed ✓
Thirty-seven		Square	
Number Novelties		A Trick for Squares	
1,089		Digit Shuffle	
1,089 Revisited		Finding the Lost Digit	
Tautonyms		Triple-digit Division	
Triple Treat		Seeing Double	
Criss Cross I		Do or Die	
Criss Cross II		Dice Dropping	
Crazy Calculator		Double Trouble	
Double up		3,087	
The Missing Eight		Cycling Digits	
Four-digit Madness		Mathemagic	
Palindromes		Calendar Magic	
Multiple Madness		Magic Months	
Think of a Number		Cool Calendar	
Give Me Five		Your Puzzle	
Diabolical Dominoes		Crazy Cards	
Dice Detective		A Math Challenge is Fun	
Pick a Card		Card Shark	
Deck Detective		Card Caper	
Create Your Own		Stepping Through Paper	
Multiples of Nine		Tied up in Knots	
Who Knows?		A Shape with a Twist I	
Mind-reader I		A Shape with a Twist II	
Mind-reader II		Missing Area I	
Number ESP		Number Words	
Missing Number		Missing Area II	
Digit Detection		Thrice Dice	
Scrambled Digits I		Mind-reading Trick	
Scrambled Digits II		Number Strips	
Double-digit Dilemma		Tricky Tables	
Ten Times		Lightning Addition	
Role Reversal		Super Sequences	
Mystery Matchbox		Mystifying Multiplication	
Calculation Shortcuts I		Number Spelling	
Calculation Shortcuts II		It all Adds up	

Magic Numbers

Several numbers, that when multiplied, appear to have magical properties. For example, the number 12,345,679. Simply ask a student to choose any single-digit number and then multiply it by 9. Next, the product is multiplied by 12,345,679. To add to the mystery of the occasion, the "mathemagician" at this point can predict the outcome prior to the completion of the final calculation.

The following example shows the trick in operation. If a student volunteer chooses 7 as the single-digit number, then 63 is produced as a result of multiplying 7 by 9 (7 x 9 = 63). The result of multiplying 63 by the magic number is 777,777,777. It should be noted that this calculation cannot be performed on an ordinary calculator with an eight-digit display, so most students are suitably impressed when the teacher ("mathemagician") predicts the answer almost instantaneously.

The product of 12,345,679 and 9 is 111,111,111. Therefore, any single-digit multiple of 9 will produce the following result *nnn,nnn,nnn*, where *n* represents the single-digit multiple of 9.

Where is the mathematics?

The mathematics involved in trick helps the students eventually develop their own magic number puzzles. Once the students appreciate the principle underlying this trick, they may like to make up some of their own, using one of the following magic numbers: 37,037, 15,873, or 8,547. To develop the trick requires recognizing the relationship between these numbers and 111,111.

Thirty-seven

☛♦ Complete the following equations:

3 x 37 =	6 x 37 =	9 x 37 =	12 x 37 =
4 x 37 =	7 x 37 =	10 x 37 =	13 x 37 =
5 x 37 =	8 x 37 =	11 x 37 =	14 x 37 =

♦ Predict what 15 x 37 equals.
♦ Use any patterns you have noticed to predict the following.

18 x 37 =   21 x 37 =   24 x 37 =

☛ Write about what you notice.

	Example	Working
☛♦ Choose a three-digit multiple of 37 from any of the numbers you have obtained so far and write it down.	518	
♦ Move the hundreds digit to the ones place and the tens digit to the hundreds place and the ones digit to the tens place to make a new number.	185	
♦ Repeat the step above.	851	

♦ Divide each of the two new numbers by 37.   185 ÷ 37 = ?   851 ÷ 37 = ?
♦ Do they divide evenly? YES/NO

- Try some different three-digit multiples of 37.
- Try adding all three numbers (518, 185 and 851) and then dividing by 37.

☛ Write about what you notice.

Number Novelties

	Example	Working
☛♦ Write down any four-digit number, without repeating a digit.	3,146	
♦ Switch the first and last digit of the number.	6,143	
♦ Subtract the smaller number from the larger.	6,143 − 3,146 2,997	
♦ Now switch the first and last digits of the difference.	7,992	
♦ Add this new number to the answer formed by the previous subtraction problem.	7,992 + 2,997	

Try • Try other four-digit numbers.

☛ Write about what you notice on the back of this page.

1,089

■♦ Choose a three-digit number where the hundreds digit is at least two more than the ones digit.

♦ Reverse the digits and then subtract the smaller number from the larger.

♦ Reverse the digits of the difference; add this number to the previous answer.

	Example	Working
	461	
	461	
	− 164	
	297	
	792	
	+ 297	
	1,089	

■♦ Repeat with other three-digit numbers and see what happens.

■ Write about what you notice.

Try
- The number 1,089 has some interesting properties. Find these products:

  1,089 × 1 =   1,089 × 6 =
  1,089 × 2 =   1,089 × 7 =
  1,089 × 3 =   1,089 × 8 =
  1,089 × 4 =   1,089 × 9 =
  1,089 × 5 =

- Discuss the pattern(s) with a friend.

1,089 Revisited

■♦ Write down any five digits in descending order.
♦ Reverse the digits.
♦ Subtract the smaller number from the larger.

	Example	Working
	97,432	
	23,479	
	97,432	
	− 23,479	
	73,953	

♦ Reverse the digits of the difference and add this new number to the answer.

	73,953	
	+ 35,937	
	109,890	

■♦ Repeat for other five-digit numbers and see what happens.

■ Write about what you notice.

Try
- Find these products:

  9 ×   1,089 =   4 ×   2,178 =
  9 ×   10,989 =   4 ×   21,978 =
  9 ×   109,989 =   4 ×   219,978 =
  9 × 1,099,989 =   4 ×   2,199,978 =
  9 ×10,999,989 =   4 × 21,999,978 =

■ Write about what you notice on the back.

Didax Educational Resources ~ www.didax.com   Magic with Math ~ 11

Tautonyms

A *tautonym* is a number where the digits are repeated twice in sequence. For example, 55, 7,272 and 397,397 are all tautonyms.

	Example	Working
☛♦ Choose any three-digit number.	295	
♦ Turn it into a tautonym (repeat it).	295,295	
♦ Divide by 7.	42,185	
♦ Divide by 11.	3,835	
♦ Divide by 13.	?	

☛♦ Repeat the steps above using another three-digit number.

♦ Does it matter whether you change the order by dividing by 13, 11 and then 7? (Use the same numbers and try it.) YES/NO

☛♦ Compare your answers with the original numbers and write about what you notice. Try to explain why the number is repeated.

Triple Treat

	Example	Working
☛♦ Choose any two-digit number.	39	
♦ Repeat it three times.	393,939	
♦ Divide by 13.	30,303	
♦ Divide by 21.	1,443	
♦ Divide by 37.	?	

☛♦ Repeat the steps above using another two-digit number.

♦ Does it matter whether you change the order by dividing by 37, 21 and then 13? (Use the same numbers and try it.) YES/NO

☛♦ Compare your answers with the original numbers and write about what you notice. Try to explain why it happens.

Try• Try writing your own number trick similar to the one above.

12 ~ **Magic with Math**   www.didax.com ~ Didax Educational Resources

Criss Cross I

	Example	Working
☛◆ Choose any number between 50 and 99.	85	
◆ Add 62 to this number.	62 + 85 = 147	
◆ Cross off the left-hand digit of the result and add it on to the ones digit.	147 ➜ 48	
◆ Subtract the result from your original number.	85 − 48	

◆ Repeat the above steps with different starting numbers.

☛ What do you notice? Try to explain what is happening.

Criss Cross II

	Example	Working
☛◆ Choose any two numbers between 50 and 100.	59 and 86	
◆ Add them.	59 + 86 ─── 145	
◆ Cross out the left-hand digit.	✗45	
◆ Add one to the remaining number.	45 + 1 = 46	
◆ Subtract the result from your added numbers.	145 − 46	

◆ Repeat the above steps using two different sets of numbers.

 • Does it work if you use 50 and 100 as your starting numbers? YES/NO

☛ What do you notice about the answers? Try to explain why it happens.

Crazy Calculator

	Example	Working
☛♦ Enter a single-digit number into your calculator.	6	
♦ Multiply it by: 3	18	
7	126	
11	1,386	
13 and then	18,018	
37.	?	

♦ Repeat the steps above starting with a different number.

☛ Write about what you notice and try to explain how it works.

☛ Create a new number trick based on the principles of the trick above.

Double up

	Example	Working
☛♦ Enter a three-digit number into your calculator.	267	
♦ Multiply by 7.	1,869	
♦ Multiply by 11.	20,559	
♦ Multiply by 13.		

♦ What happened? Discuss the results with a partner.

☛♦ Repeat the steps above with other three-digit numbers.

Try
- Try three-digit numbers where the digits are all the same; e.g. 222, 555, 888.
- Try three-digit numbers with trailing zeros; e.g. 100, 400, 700.
- Does it matter when you change the order and multiply by 13, 11 and then 7?

☛ Write about what you notice and try to explain why it happens.

The Missing Eight

	Example	Working
☛◆ Write down the numbers 1 to 9, but leave out the 8.	12,345,679	
◆ Choose one of the eight numbers.	4	
◆ Multiply this number by 9.	9 x 4 = 36	
◆ Multiply your answer by 12,345,679.		
☛◆ Repeat the same procedure, but this time choose a different number to multiply by 9.		

Try
- Predict what the answer will be for any multiple of 9.
- Try to get every number from 1 to 9 excluding 8.

☛ What do you notice about your answers? Try to explain why the 8 has been left out.

Four-digit Madness

	Example	Working
☛◆ Write down any four-digit number where all four digits are different.	5,643	
◆ Rearrange the four digits to produce:		
(a) the largest number you can	6,543	
(b) the smallest number you can	3,456	
◆ Subtract the smaller number from the larger.	6,543 − 3,456	
◆ Now you should have a new number.	3,087	
◆ Do the same to this number.	8,730 − 378 8,352	
☛◆ Continue doing this until you reach 6,174. (It shouldn't take more than seven tries to reach 6,174.)	8,532 − 2,358 6,174	

Try
- Find starting numbers that produce 6,174 in one, two, three and four tries.
- What happens if you start with 6,174?
- Will the trick still work if you start with a number where all four digits are not different? Try 1,355, 5,544 and 2,333.

☛ Write about what you notice and try to explain why it happens.

Didax Educational Resources ~ www.didax.com   Magic with Math ~ 15

Palindromes

A *palindrome* is a word or number that reads the same both forwards and backwards; for example, mom, dad, noon, radar, 99, 323, 747, etc.
There are even palindromic sentences.

MADAM I'M ADAM.
A MAN, A PLAN, A CANAL, PANAMA.

You can produce your own palindromes using the following procedures:

	Examples	Working
☛♦ Select a number.	18	
♦ Reverse the digits.	81	
♦ Add the numbers.	99	
♦ In one step (reversing and adding) we have formed a palindromic number. The number 28 requires two steps.	28 + 82 110	110 + 011 121

You will find most numbers form a palindromic number in very few steps.

- Try these: 123, 632, 458, 184, 291 and 79.
- Use some of your own numbers.

Decimal palindromes can be produced using the following procedure:

☛♦ Select a decimal number.	5.13	
♦ Reverse the digits.	31.5	
♦ Add the numbers.	36.63	
♦ It is important that the decimal point be placed in the correct position, otherwise a palindrome is not formed. Some numbers require more than one step.	21.9 + 9.12 31.02	31.02 + 20.13 51.15

You will find most numbers form a palindromic number in very few steps.

- Try these numbers: 15.4, 25.4, 16.7, 64.7 and 98.3
- Try starting with a palindrome; e.g. 21.12, 16.61, 17.71 and 51.15.

Multiple Madness

	Example	Working
☛♦ Choose any multiple of 7.	(5 x 7)	35
♦ Multiply this number by 15,873.		
♦ Record the result.		
☛♦ Choose any multiple of 13.	(3 x 13) 39	
♦ Multiply this number by 8,547.		
♦ Record the result.		

- Now try other multiples of 7, then other multiples of 13.

☛ Discuss what you notice with a partner.

Explanations

Thirty-seven – page 10

3 × 37 = 111	4 × 37 = 148	5 × 37 = 185
6 × 37 = 222	7 × 37 = 259	8 × 37 = 296
9 × 37 = 333	10 × 37 = 370	11 × 37 = 407
12 × 37 = 444	13 × 37 = 481	14 × 37 = 518
15 × 37 = 555	18 × 37 = 666	21 × 37 = 777
24 × 37 = 888		

This pattern occurs because 3a × 37 = 111a (where a is a number from 1 to 27).

When the digits in a three-digit dividend are rearranged, according to the directions, no remainder occurs when divided by 37.

Number Novelties – page 10

Let a, b, c and d represent the digits of the starting number. The four-digit number would be represented by:

$$1{,}000a + 100b + 10c + d$$

Switching the digits and subtracting the smaller number from the larger gives:

$1{,}000a + 100b + 10c + d - (1{,}000d + 100b + 10c + d)$

$= 999a - 999d$

or $999(a - d)$

If the difference between the first and last digits is 1, the answer will always be 1998 (e.g. 999 + 999). The result for all other values will be 10,989.

1,089 – page 11

The same pattern—900 + 180 + 9—occurs as a result of the restrictions set at the beginning of the puzzle.

The process that leads to the answer of 1,089 may be represented algebraically. If a, b and c represent the original digits of the number, then we get:

$100a + 10b + c - (100c + 10b + a) = 99a - 99c$
$\qquad\qquad\qquad\qquad\qquad\quad$ or $99(a - c)$

We know that a and c can only represent single-digit numbers and that they cannot be equal, due to the restrictions set at the beginning of the puzzle. Therefore, the differences are restricted to multiples of 99; e.g. 198, 297, 396, 495, 594, 693, 792 and 891.

When the numbers are reversed and added, the following combinations:

099	198	297	396	495
+ 990	+ 891	+ 792	+ 693	+ 594
1,089	1,089	1,089	1,089	1,089

and vice versa

Similar reasoning applies to the number 1,089.

1,089 × 1 = 1,089	1,089 × 2 = 2,178
1,089 × 3 = 3,267	1,089 × 4 = 4,356
1,089 × 5 = 5,445	1,089 × 6 = 6,534
1,089 × 7 = 7,623	1,089 × 8 = 8,712
1,089 × 9 = 9,801	

There are several patterns:

The thousands and hundreds digits both increase by one while the tens and ones digits both decrease by one. If you add the first two digits and the last two digits, you always end up with 99. Going down the list of answers, each number comes up again later, but in reverse.

1,089 Revisited – page 11

The final sum always remains the same – 109,890.

There are several patterns in the multiplications:

9 × 1,089 =	9,801
9 × 10,989 =	98,901
9 × 109,989 =	989,901
9 × 1,099,989 =	9,899,901
9 × 10,999,989 =	98,999,901

4 × 2,178 =	8,712
4 × 21,978 =	87,912
4 × 219,978 =	879,912
4 × 2,199,978 =	8,799,912
4 × 21,999,978 =	87,999,912

Note: 2,178 is 2 × 1,089; therefore 4 × 2,178 is the same as 4 × 2 (or 8) × 1,089.

Tautonyms – page 12

Following this procedure will always produce the original three-digit number.

Dividing by 7, 11 and then 13 is the same as dividing by 1,001. By repeating your three digits to produce a tautonym you have effectively multiplied your three-digit number by 1,001. Dividing by 1,001 (7 × 11 × 13) returns you to you original number. The order in which the division is carried out makes no difference to the final answer.

Triple Treat – page 12

You finish with the number you started with.

This will always happen because dividing by 13, 21 and 37 is the same as dividing by 10,101. If you multiply any two-digit number by 10,101 you will notice that the two-digit number is repeated three times to produce a six-digit number; e.g. 39 × 10,101 = 393,939. The trick relies on reversing this procedure. You may like to write your own number trick based on multiplying a two-digit number by 1,010,101. The order in which the division is carried out makes no difference to the final answer.

Explanations

Criss Cross I – page 13

The result will always be 37 because when you cross off the leftmost digit of the result, which is always 1**, and add it on to the ones digit, you are effectively subtracting 99. Earlier, the instruction was to add 62. The difference between 99 and 62 is 37.

If you wish to alter the result, simply change the number that you add in the second step. The difference between this number and 99 will be the answer you get at the end of the puzzle.

Criss Cross II – page 13

This trick is very similar to Criss Cross I. The starting number has to be between 50 and 100, thereby ensuring that the result will be between 100 and 200. Crossing out the leftmost digit and adding one is the same as subtracting 99.

When the remaining part is subtracted from the original sum it has to leave 99 as the answer.

Crazy Calculator – page 14

$3a \times 7 \times 11 \times 13 \times 37 = 111{,}111a$ (where a is a single-digit number). In every case, the number chosen will produce six digits of that particular number.

Double up – page 14

Multiplying the original three-digit number by 7, 11 and 13 is the same as multiplying by 1,001. Multiplying by 1,000 moves the three-digit number three decimal places to the left, while multiplying by 1 ensures that the same digits will be repeated in the hundreds, tens and ones places. The order in which the multiplication is carried out makes no difference to the final answer.

There are several variations to this trick. For example, you can reduce the number of steps by asking your students to multiply by 7 and then by 143 (11 x 13) or by 11 and 91 (7 x 13). Another variation involves starting with a two-digit number and multiplying by 13, 31 and 37. This is the same as multiplying by 10,101, and therefore, the same pair of digits is repeated twice; e.g. 64 x 13 x 21 x 37 = 646,464. This trick may also be varied by changing the order in which the multiplication occurs. Both tricks may be altered to incorporate division rather than multiplication.

The Missing Eight – page 15

The answer is made up entirely of the number chosen in step 2.

This is because 9 x 12,345,679 = 111,111,111; therefore, 9a x 12,345,679 = 111,111,111 (where a is a number 1 to 9, excluding 8).

When multiples of 9 are multiplied by 12,345,679 the answer will consist entirely of the digit corresponding to the multiple of 9 that was chosen.

Four-digit Madness – page 15

6,174 gives
$$\begin{array}{r} 7{,}641 \\ -\ 1{,}467 \\ \hline 6{,}174 \end{array}$$ in one try

1,355 gives 6,174 in two steps.
5,544 gives 6,174 in four steps.
2,333 doesn't work.

However, some numbers with three repeating digits do work. For example, 1,777 gives 6,174 in six steps. Obviously, a number with all four repeating digits would not work.

Palindromes – page 16

The difficulty level of this exercise may be altered by changing the numbers. However, the size of the numbers you choose may have little to do with the number of steps it takes to produce a palindrome.

123 (1 step), 632 (1 step), 458 (2 steps), 184 (3 steps), 291 (4 steps), 79 (6 steps)

When you begin with a palindrome, you should end up with a palindrome.

15.4 (1 step), 25.4 (1 step), 16.7 (2 steps),
64.7 (3 steps), 98.3 (2 steps)
21.12 (1 step), 16.61 (2 steps), 17.71 (2 steps),
51.15 (2 steps)

Multiple Madness – page 16

15,873 x 7 = 111,111
15,873 x 7 x 5 = 555,555
15,873 x 7 x n = nnn,nnn (where n is a single-digit number)

8,547 x 13 = 111,111
8,547 x 13 x 3 = 333,333
8,547 x 13 x n = nnn,nnn (where n is a single-digit number)

In order for this problem to work, you must choose a single-digit number. In every case, the number chosen will produce six digits of that particular number.

Notes

"Think of a Number" Tricks

"Think of a Number" Tricks may be found in most teachers' "bag of tricks." The teacher guides the students through a series of mental calculations that either end up with the starting number or a number stated at the beginning of the trick. Not only do these tricks provide students with the motivation to complete a series of mental calculations, but they can also be used to introduce some simple algebra. The link between the trick and the algebra may be seen in the following example. Once students have seen how this type of "mathemagical trick" is created, they can design some of their own. This approach provides a simple introduction to algebra. Students can be challenged to modify the example below or create one of their own that ends up with a predetermined answer. For example, adding the instruction, "subtract the number you first thought of" removes the original number anywhere mentioned in the set of instructions.

Instructions		Example	Algebraic explanation
1.	Choose a number.	6	x
2.	Double it.	12	2x
3.	Add 10.	22	2x +10
4.	Triple it.	66	6x + 30
5.	Subtract 30.	36	6x
6.	Divide by 6.	6	x

The example below illustrates how a "think of a number" puzzle may be designed to produce a predetermined answer.

Instructions		Example	Algebraic explanation
1.	Choose a number.	6	x
2.	Add 20.	26	x + 20
3.	Double it.	52	2x +40
4.	Subtract 10.	42	2x + 30
5.	Halve it.	21	x + 15
6.	Subtract first number.	15	15

One way to start students thinking about puzzles of this type is to give them one that works (e.g. the one shown above) and ask them to change the puzzle to finish on a different number. Changing step 2 or step 4 will change the final number.

Think of a Number

Trick 1	Example	Working	Trick 2	Example	Working
☛◆ Think of a number.	20		☛◆ Think of a number.	15	
◆ Triple it (multiply by 3).	60		◆ Double it.	30	
◆ Add 10.	70		◆ Add 100.	130	
◆ Double it.	140		◆ Halve it.	65	
◆ Subtract 14.	126		◆ Subtract 29.	36	
◆ Divide by 6.	21		◆ Subtract your original number.	?	
◆ Subtract 1.	?				

☛ Repeat both of the tricks above with several different numbers. Record your findings below and try to explain how each works.

Trick 1

Trick 2

Give Me Five

	Example	Working
☛◆ Choose any number.	23	
◆ Add the next counting number to the original number.	23 + 24 = 47	
◆ Add 9.	47 + 9 = 56	
◆ Divide by 2.	56 ÷ 2 = 28	
◆ Subtract your original number.	28 − 23 = ?	

☛◆ Repeat the steps above using different starting numbers.

 Try
- Try using a single-digit number.
- Try using a three-digit number.

☛ Write about what you notice. Try to explain why it happens.

Diabolical Dominoes

- ◆ Work with a partner. Keep this sheet hidden from him/her while you work your "magic" trick.
- ◆ Ask your partner to choose any domino piece while you turn your back, and then follow this procedure:
  - Multiply one of the numbers shown on the domino by 5.
  - Add 6 to the result.
  - Double your answer.
  - Add the number on the other half of the domino.
  - Ask your partner to tell you his/her final number.
- ◆ Use your "Secret Solution" to reveal the numbers on the domino without ever having seen them.

Secret Solution

Subtract 12 to find the two numbers that were on the original domino. One number will be the tens digit, the other number will be the ones digit.

➡ Explain how you and your partner think this trick works.

- Use two dice instead of dominoes to get the starting numbers.
- Make up a different set of instructions; for example:
  - Multiply one of the numbers by 5.
  - Add 7. Double it.
  - Add the number shown on the other die.
  - Subtract 14 to find the starting numbers.

Dice Detective

- ◆ Work with a partner. Keep this sheet hidden from him/her while you work your "magic" tricks.

Trick 1

- ◆ Ask your partner to roll two different colored dice (e.g. black and white) while you turn your back, and then follow this procedure:
  - Double the number on the black die.
  - Add 5 to the answer.
  - Multiply the result by 5.
  - Add the number showing on the white die to the answer.
  - Ask your partner to tell you the result.
- ◆ Use your "Secret Solution" to reveal the numbers on the two dice.

Secret Solution

To reveal the numbers, subtract 25 from the result. The first number is the one showing on the black die and the second number the one on the white die.

Trick 2

- ◆ Ask your partner to roll two different colored dice (e.g. black and white) while you turn your back, and then follow this procedure:
  - Multiply the number on the black die by 5.
  - Add 7 to the answer.
  - Double the result.
  - Add the number showing on the white die to the answer.
  - Ask your partner to tell you the result.
- ◆ Use your "Secret Solution" to reveal the numbers on the two dice.

Secret Solution

To reveal the numbers, subtract 14 from the result. The first number is the one showing on the black die and the second number the one on the white die.

Pick a Card

👉♦ Work with a partner. Keep this sheet hidden from him/her while you work your "magic" trick.

	Example	Working
♦ Ask your partner to choose any card from your deck of cards while you turn your back. Follow this procedure:	9 of clubs	
• Double the face value.*	18	
• Add 4.	22	
• Multiply by 5.	110	
• Add the suit value (in this case 2 – see below).	112	

2 for a club   4 for a diamond   6 for a heart   8 for a spade

• Ask your partner to tell you his/her final number.
♦ Use your "Secret Solution" to reveal the card without ever having seen it.

Secret Solution
To identify the card subtract 20 from the result.
The tens digit gives the face value of the card and the ones digit the suit, e.g. 112 – 20 = 92 or 9 of clubs.

*Ace = 1
Jack = 11
Queen = 12
King = 13

👉 Try to explain how you and your partner think this trick works on the back.

Deck Detective

👉♦ Work with a partner. Keep this sheet hidden from him/her while you work your "magic" trick.
  ♦ Separate the cards from Ace to 6 from a standard deck of cards; 24 cards should remain.
  ♦ Turn your back and ask your partner to choose three cards from the deck and lay them out to form a three-digit number, e.g. 5 6 2.
  ♦ Ask your partner to follow this procedure:
   • Double the value shown on the first card.
   • Add 5 to the answer.
   • Multiply the result by 5.
   • Add the number showing on the second card to the answer.
   • Multiply the result by 10.
   • Add the value shown on the third card to the answer.
   • Ask your partner to tell your the result.
  ♦ Use your "Secret Solution" to reveal the numbers on the three cards without ever having seen them.

Secret Solution
To work out the values on the original cards subtract 250 from the result. The answer represents the values of each of the three cards.

Create Your Own

- ♦ Create your own "think of a number puzzle."
- ♦ There are two types:
  - Steps are followed that end with the original number chosen (or with a number that helps you to work out the original number).
  - Steps that lead to a specific number at the end.
- ♦ To be able to create a number trick, you need to know a little algebra. Here are some helpful hints:

Instruction	Example	Algebra
think of a number	13	x
double the number	26	$2x$
triple the number	78	$6x$
add 6	84	$6x + 6$
subtract 2	82	$6x + 4$
double the result	164	$2(6x + 4) = 12x + 8$

If you add or subtract a number and then multiply, remember to multiply everything; for example, if you were to double $x - 2$ you would have $2x - 4$.

- ♦ Before you start, read through these tricks and make sure you understand how they work (or have the solutions nearby so that you can follow them):
  - Think of a Number
  - Give Me Five
  - Dice Detectives
  - Deck Detectives.

Your Puzzle

Secret Solution

Explanations

Think of a Number – page 21

Trick 1
This trick will always end up with the starting number. A little algebra shows why.

Think of a number.	n
Triple it.	$3n$
Add 10.	$3n + 10$
Double it.	$6n + 20$
Subtract 14.	$6n + 6$
Divide by 6.	$n + 1$
Subtract 1.	n

Trick 2
This trick will always end up with the same number (21). A little algebra shows why.

Think of a number.	n
Double it.	$2n$
Add 100.	$2n + 100$
Halve it.	$n + 50$
Subtract 29.	$n + 21$
Subtract original number	21

The final number may be altered easily by changing the number you subtract in step 5. You may subtract any number from 1 to 49. To work out your first number, subtract the number you use from 50.

You may also wish to alter the number added in step 3. Any even number may be substituted for 100. To work out your final number, halve the number you add and subtract the number used in step 5.

Give Me Five – page 21

Choose a number.	n
Add the next counting number	$n + (n + 1)$
or	$2n + 1$
Add 9.	$2n + 10$
Divide by 2.	$n + 5$
Subtract the original number.	5

This trick may be altered by substituting another odd number for 9. The answer will always be one-half of one more than the odd number added.

Diabolic Dominoes – page 22

This mathematical activity is simply a variation of **Think of a Number**. Instead of the student supplying the number, the domino (or die) does.

Let n be the starting number.

Multiply by 5.	$5n$
Add 6.	$5n + 6$
Double.	$10n + 12$
Add the second domino (x).	$10n + x + 12$
Subtract 12.	$10n + x$

As dominoes can only contain single-digit numbers, the first number will always be in the tens place. (The instructions to multiply by 5 and 2 ensure this.) The second number will always be in the ones place.

Dice Detective – page 22

Trick 1
Let a and b represent the values shown on the two dice. Following the instructions produces $2a$, $2a + 5$, $10a + 25$, $10a + 25 + b$, $10a + b$.

(Note: The 10 in $10a + b$ denotes the first number being in the "tens place"; e.g. 32, $10a = 30$ so $a = 3$.)

Trick 2
Let a and b represent the values shown on the two dice. Following the instructions produces $5a$, $5a + 7$, $10a + 14$, $10a + 14 + b$, $10a + b$.

Pick a Card – page 23

Pick a card.	9 of clubs	a
Double the face value.	18	$2a$
Add 4.	22	$2a + 4$
Multiply by 5.	110	$10a + 20$
Add suit value (club = 2).	112	$10a + 20 + $ suit value
Subtract 20.	92	$10a + $ suit value

The trick may be altered by changing the values added for the suit. These values, however, should always be single digits to ensure that the suit value ends up in the ones place; e.g. 1 for a club, 2 for a diamond, 3 for a heart, 4 for a spade.

Deck Detective – page 23

Let a, b and c represent the values shown on the three cards chosen. Following the instructions produces:

$2a$, $2a + 5$, $10a + 25$, $10a + 25 + b$, $100a + 250 + 10b$, $100a + 250 + 10b + c$, $100a + 10b + c$

Create Your Own – page 24

Teacher check

Notes

Properties of 9

A simple trick involving nines is outlined below. Prior to starting this trick write "9" on a piece of paper and seal it in an envelope. Ask a student to think of a three-digit number, e.g. 899. Next, ask the student to multiply the number by 9 (giving 8,091). Instruct the student to add all the digits in the answer until a single digit answer is reached. (8 + 0 + 9 + 1 = 18, 1 + 8 = 9). Ask the student to open the envelope to reveal the predicted answer.

Some students may be aware that when you multiply a number by 9 and add all the digits of the result until a single-digit number is formed, the result will always be 9.

For example: 137 x 9 = 1,233
1 + 2 + 3 + 3 = 9

This occurs because we use a base 10 number system. This property may be used to check the results of a calculation, using a method known as "casting out nines." The same property may be used as the basis for some mathemagic.

The tricks in this next section all rely upon the magical properties of the number 9 and in some cases a little algebra to produce some amazing results.

Multiples of Nine

	Example	Working
☛♦ Write down any two-digit number.	61	
♦ Reverse the digits.	16	
♦ Find the difference (in other words subtract the smaller number from the greater).	61 – 16 = 45	
♦ Divide by 9.	45 ÷ 9 = 5	
☛♦ Repeat the steps above with other two-digit numbers.		

- Do you always get multiples of 9? YES/NO
- Can you get all the multiples of 9 from 9 to 81? YES/NO

☛ Write about what you notice and try to explain why it happens.

☛ Can you find any connection between the number you start with and the multiple of 9 you end up with?

Who Knows?

	Example	Working
☛♦ Choose a three-digit number. (Do not choose one which uses the same digit three times.)	437	
♦ Use the same three digits to form the largest three-digit number you can.	743	
♦ Now form the smallest three-digit number you can.	347	
♦ Find the difference between the larger and smaller numbers.	743 – 347 = 396	
♦ Repeat the process using the three digits from your new number.	963 – 369 = 594	
♦ Repeat the process one more time using the three digits from your new number.	954 – 459	

♦ Repeat the same procedure using other three-digit numbers.

☛ What do you notice about the answers?

- Try using 747. What do you notice?

Mind-reader I

☛♦ Work with a partner. Keep this sheet hidden from him/her while you work your "magic" trick.

	Example	Working
♦ Ask your partner to secretly write down any single-digit number, then follow this procedure:	6	
• Multiply this digit by 10.	60	
• Add the original digit.	66	
• Multiply the result by 9.	594	
• Multiply by 11.	6,534	

♦ Ask your partner to tell you the last digit in the result.
♦ Use your "Secret Solution" to reveal the other three digits.

Secret Solution

The other three digits are found in this way:
1. The tens digit is one less than the ones digit.
2. The hundreds digit is the difference between 9 and the ones digit.
3. The thousands digit is the difference between 9 and the tens digit.

☛♦ Discuss with your partner how you think this trick works.

Mind-reader II

☛♦ Work with a partner. Keep this sheet hidden from him/her while you work your "magic" trick.

	Example	Working
♦ Ask your partner to secretly write down any four-digit number, where all the digits are different, then follow this procedure:		
• Reverse the order of the digits.	1,537	
• Take the smaller number away from the larger number.	7,351 − 1,537 = 5,814	
• Multiply the result by any number from 2 to 9,999.	5,814 × 3 = 17,442	
• Cross off any one digit from your answer except 0 or 9.	1~~7~~,4 4 2	
• Now ask your partner to tell you his/her remaining digits. (Note, the order is not important.)	1,4 4 2	

♦ Using your "Secret Solution," offer to "read" his/her mind to find out the number that was crossed off.

Secret Solution

To work out which digit was crossed off, simply add all the digits together and subtract your answer from the next multiple of 9. Using the above example, the digits 1 + 4 + 4 + 2 = 11. The next multiple of 9 is 18; therefore, the missing digit is 7.

Number ESP

These two tricks are variations on a theme. Try the two-digit one first and then try the more difficult three-digit version.

☛♦ Work with a partner. Keep this sheet hidden from him/her while you work your "magic" tricks.

Two-digit Trick

	Example	Working
☛♦ Ask your partner to secretly write down a two-digit number, then follow this procedure:
- Reverse the number.
- Subtract the smaller number from the larger.

```
   91
 - 19
 ----
   72
```
Example: 19, 91, 91 − 19 = 72

- Ask your partner to tell you the right-hand digit of the result. → 2

♦ Use your "Secret Solution" to tell him/her what the final result was.

Secret Solution

Supply the answer by subtracting the given digit from 9 to determine the tens digit.

Three-digit trick

	Example	Working
☛♦ Ask your partner to secretly write down a three-digit number, then follow this procedure:
- Reverse the number.
- Subtract the smaller number from the larger.

```
   611
 - 116
 -----
   495
```
Example: 611, 116, 611 − 116 = 495

- Ask your partner to tell you the right-hand digit of the result. → 5

♦ Use your "Secret Solution" to tell him/her what the final result was.

Secret Solution

Here is how to do it:
- The tens digit will always be 9, unless the right-hand digit is 0, in which case the answer will be 0.
- If the student tells you the right-hand digit is 9, then the answer is 99.
- If the right-hand digit is a digit other than 9 or 0, then subtract that number from 9 to arrive at the hundreds digit. Using the example above, if the student told you the last digit was 5, you would know that the tens digit is 9 and the hundreds digit is 4 (because 9 − 5 = 4), so the number would be 495.

☛♦ With your partner, discuss what you notice and try to explain why it happens.

Missing Number

☛◆ Work with a partner. Keep this sheet hidden from him/her while you work your "magic" trick.

	Example	Working

◆ Ask your partner to secretly write down any four-digit number where all the digits are different, then follow this procedure: 6,784
  • Add all the digits.   6 + 7 + 8 + 4 = 25
  • Keep adding the digits of your sum until you reach a single-digit number.   2 + 5 = 7
  • Cross out one of the original digits.   67̶8̶4
  • Subtract the reduced number from the remaining three-digit number.

      674
    −   7
    ―――
      667

  • Ask your partner to tell you the result.
◆ Using the "Secret Solution" announce the number that was crossed out.

Secret Solution

To find the missing number, add the digits of your final answer together and keep adding until you reach a single-digit number. Subtract this number from 9 to find the missing number.

☛◆ Discuss with your partner how you think this trick works.

Digit Detection

☛◆ Work with a partner. Keep this sheet hidden from him/her while you work your "magic" trick.

	Example	Working

◆ Ask your partner to secretly write down any three-digit number where all the digits are different, then follow this procedure:
  • Write down the three-digit number.   617
  • Add the digits.   6 + 1 + 7 = 14
  • Subtract the sum of the digits from your three-digit number.   617 − 14 = 603
  • Ask your partner to tell you the first two digits of his/her result.   6 and 0
◆ Use your "Secret Solution" to tell him/her the result.

Secret Solution

To find the last digit of the answer, add the two digits given to you and subtract the sum from the next multiple of 9.
For example:
supplied digits   6 and 0
add them together   6 + 0 = 6
subtract from the next multiple of 9   9 − 6 = 3
6 and 0 were supplied so the number is: 603

☛ ◆ Discuss with your partner how you think this trick works.

Scrambled Digits I

	Example	Working
☞♦ Choose any five-digit number.	56,134	
♦ Scramble the digits.	34,165	
♦ Subtract the smaller number from the larger.	56,134 − 34,165 21,969	
♦ Add the digits that form the answer until you reach a single digit.	2 + 1 + 9 + 6 + 9 = 27 2 + 7 = 9	

♦ Repeat the steps above using different five-digit numbers.

> **Try** • Try starting with 2-, 3-, 4- and 6-digit numbers and write down what happens.

☞ Write about what you notice.

Scrambled Digits II

☞♦ Work with a partner. Keep this sheet hidden from him/her while you work your "magic" trick.

	Example	Working
♦ Ask your partner to secretly write down any six-digit number where all the digits are different, then follow this procedure:	458,613	
• Add the digits in the number.	4 + 5 + 8 + 6 + 1 + 3 = 27	
• Subtract this amount from your original number.	458,613 − 27 458,586	
• Scramble the digits of the resulting number.	558,684	
• Add 25.	558,684 + 25 558,709	
• Cross out any digit except a zero.	5 5 8 ̷7 0 9	
• Find the sum of the remaining digits.	5 + 5 + 8 + 0 + 9 = 27	

♦ Ask your partner to write the result and pass it to you.
♦ Using the "Secret Solution," tell your partner which digit was crossed out.

Secret Solution

> To find the missing digit, simply subtract 7 from the final digit sum and then subtract this amount from the next highest multiple of 9.

☞♦ With your partner, discuss what you notice and try to explain why it happens.

Double-Digit Dilemma

	Example	Working
☛ ◆ Choose a two-digit number where no digit is repeated.	27	
◆ Reverse the digits.	72	
◆ Find the difference of the two numbers.	72 − 27 45	
◆ Find the difference between your original two digits. (Take the smaller number away from the larger.)	7 − 2 = 5	
◆ Divide your first answer by this difference.	45 ÷ 5 = ?	
◆ Repeat the steps above with different two-digit numbers.		

☛ What do you notice? Try to explain why it happens.

Ten Times

	Example	Working
☛ ◆ Randomly choose a number. (Note: The number can be made up of any number of digits and digits may be repeated.)	4,751	
◆ Multiply your number by 10.	47,510	
◆ Subtract the original number from the product.	47,510 − 4,751 42,759	
◆ Use a calculator to divide this answer by your original number.	42,759 ÷ 4,751 = ?	
◆ Repeat the steps above with some different starting numbers.		

Try • Try single-digit, two-digit, three-digit and four-digit numbers.

☛ Write about what you notice. Try to explain why it happens.

☛ With a partner, create a trick similar to the one above.

Role Reversal

	Example	Working
☛♦ Choose any two-digit number where no digit is repeated.	63	
♦ Reverse the digits.	36	
♦ Subtract the smaller number from the larger.	63 − 36 ―― 27	
♦ Repeat the process using the new number that is formed.	72 − 27 ―― 45	
♦ Continue until you reach a single-digit answer.	54 − 45	

♦ Repeat the steps above for these numbers:
65 (1 step), 71 (2 steps), 93 (2 steps), 74 (3 steps), 26 (4 steps), 57 (5 steps).

☛♦ What answer do you always reach?

☛♦ Find a pattern to determine whether a starting number will take just one step to reach an answer. Write your findings below.

Mystery Matchbox

☛♦ Work with a partner. Keep this sheet hidden from him/her while you work your "magic" trick.

	Example	Working
♦ Ask your partner to secretly count the number of matches in a matchbox.	48	
♦ Next, ask your partner to add the two digits of the number.	4 + 8 = 12	
♦ Instruct your partner to remove this number of matches from the matchbox.	48 − 12 = 36	
♦ Ask your partner to hand the matchbox to you.		
♦ Shake the matchbox and use the "Secret Solution" to reveal the number of matches in the box.		

Secret Solution

If the your partner has carried out these steps correctly, then a multiple of 9 is always left. All you have to do is guess from the feel of the matchbox whether it contains 45, 36, 27, 18 or 9 matches.

To eliminate the possibility that you may have rigged the box, you could get your partner to secretly remove some matches from the box before starting to count them.

Explanations

Multiples of Nine – page 28

You always get a multiple of 9, and you can get all of the multiples of 9.

Simply find the difference between the tens digit and the ones digit in your starting number to reveal the multiple of 9 that you will end up with.
For example, 61, 6 – 1 = 5. The answer was 45 and 9 x 5 = 45.

Who Knows? – page 28

If we let the digits be represented by a, b and c and assume a>b and b>c
(100a + 10b + c) – (100c + 10b + a) = 99a – 99c or 99(a – c).

The difference between a and c will always be at least 2 and never more than 8 so the multiples formed will be 198, 297, 396, 495, 594, 693, 792.

In each case, the largest digit is 9. When the largest three-digit number is formed the 9 will always end up in the hundreds position. The 9 will always be in the ones place when the smallest number is formed.

e.g.
```
    981         972         963         954
  – 189       – 279       – 369       – 459
    792 leads to 693       594 leads to 495
```

The rest form the same pattern.
The result is always 495.

Mind-reader I – page 29

Let n denote the chosen digit. The steps produce (10n + n) x 9 x 11 = 1,089n.

This number is always divisible by 9 and 11. This means that the sum of the digits that make up the final result will equal 9 or a multiple of 9.

1,089 x n always produces a number where the tens digit is one less than the ones digit; e.g. 4 x 1,089 = 4,356—the ones digit and the hundreds digit will always add to nine. The zero in 1,089 ensures that no digits are carried over into the thousands place; the thousands digit will be the same as the single digit multiplier (n), because 1 x n = n. The thousands digit and the tens digit add to make 9.

Mind-reader II – page 29

Following the instructions gives:
1,000a + 100b + 10c + d – (1,000d + 100c + 10b + a) = 999a + 90b – 90c – 999d
or 9(111a + 10b – 10c – 111d).
Subtracting the smaller number from the larger number produces a multiple of 9. Multiplying this value by any number from 2 to 9,999 still produces a multiple of 9. Therefore, to find the missing digit, all you have to do is subtract the total of the given digits from the next multiple of 9.

Number ESP – page 30

Why it works:
- The tens digit will always be 9, unless the right-hand digit is 0, in which case the answer will be 0.
- If the student tells you the ones digit is 9, then the answer is 99.
- If the ones digit is a digit other than 9 or 0, then subtract that number from 9 to arrive at the hundreds digit.

Using the example given, if the student told you the last digit was 5, you would know that the tens digit is 9 and the hundreds digit is 4 (because 9 – 5 = 4) so the number would be 495.

Missing Number – page 31

To find the missing number, reduce the answer by adding the digits until a single digit is produced.
e.g. 6 + 6 + 7 = 19
      1 + 9 = 10
      1 + 0 = 1

Subtract the reduced one-digit number from 9 (e.g. 9 – 1 = 8) to reveal the missing number.

Digit Detection – page 31

To find the last digit of the answer, add the two digits given to you and subtract the sum from the next multiple of 9.
For example:

supplied digits	6 and 0
add them together	6 + 0 = 6
subtract from the next multiple of 9	9 – 6 = 3
six and zero were supplied so the number is:	603

Explanations

Scrambled Digits I – page 32

The result is always 9 no matter what numbers you use.

Consider the simplest case of "Scrambled Digits" which involves a two-digit number ab. The first three instructions lead to the following:

$(10a + b) - (10b + a) = 9a - 9b$ (assuming $a > b$)
or $9(a - b)$

The number formed will be a multiple of 9; hence, the digits will always add up to make 9.

Scrambled Digits II – page 32

Subtracting the sum of the digits in the original number results in digits that add up to an exact multiple of 9. Adding 25 to the digit sum of the scrambled answer produces a sum that is 7 more than the sum of the original digits, an exact multiple of 9. Therefore, to find the missing digit, simply subtract 7 from the final digit sum and then subtract this amount from the next highest multiple of 9.

Variation: Instead of adding 25, you may use any other number. When working out the missing digit, instead of subtracting 7 from the final digit sum, you need to subtract the digit sum of the number you added earlier (e.g. if you added 32, you would subtract 5 (3 + 2 = 5)) from the digit sum and then subtract that number from the next highest multiple of 9.

Double-digit Dilemma – page 33

This trick is similar to **Scrambled Digits**.

You will always end up with 9 as your answer. In other words, a multiple of 9 is always produced.

If the two digits are represented by a and b then the steps become:

$(10a + b) - (10b + a) = 9a - 9b$ or $9(a - b)$.

The difference between the two original digits is $a - b$, therefore dividing by $a - b$ will always leave 9 as the answer.

Ten Times – page 33

This process may be easily altered by changing the second instruction—"Multiply by 10"—to "Add a zero at the end." Both instructions have the same effect—moving the number one decimal place to the left.

Why it works:

let the random number be	a
Multiplying by 10 gives	$10a$
Subtracting the original number	$10a - a = 9a$
Dividing by the original number	$9a \div a = 9$

Role Reversal – page 34

If we call our starting digits x and y, then the two-digit number would be $10x + y$. Reversing the digits, we get $10y + x$.

$(10x + y) - (10y + x)$ leaves $9x - 9y$ or $9(x - y)$

This means that we will always end up with a multiple of 9 after the first subtraction. Repeating the process eventually leads us to the point where the difference between the digits x and y is 1, and $9(x - y) = 9$.

Mystery Matchbox – page 34

If the partner has carried out these steps correctly, then a multiple of 9 is always left. If the starting number is represented by $10a + b$ then subtracting the sum of the digits, $a + b$ leaves $9a$ which must produce a multiple of 9.

Notes

Algebra-based Tricks

The following tricks may be explained with the use of some simple algebra.

They are ideal for:

- baffling a younger audience; or
- developing algebra with older students.

Once older students know the algebra behind the tricks, they can try writing some of their own. The algebra involved in the tricks contained in this section builds on the algebra concepts in the "Think of a Number" section. Some of tricks involve multiplying binomials, while others require students to relate place value concepts and algebra. For example, when a two-digit number is formed from the digits a and b, the result may be either $10a + b$ or $10b + a$.

A simple trick involving squaring numbers is illustrated below.

One Up, One Down

$8^2 = 8 \times 8$ or 64

If the first number is increased by one, and the second number is decreased by one, the following results:

$9 \times 7 = 63$.

Test some examples. What do you notice?

This trick is based on simple algebra.

The first instruction is a^2

The second gives $(a + 1)(a - 1)$ which, when multiplied, gives $a^2 + a - a - 1$ or $a^2 - 1$. An understanding of this principle may be used as a basis of a "mathemagic" trick.

The multiplication shortcut that follows is based on the difference of squares; i.e. $(a - b)(a + b) = a^2 - b^2$.

(Note: This comment refers to the multiplication shortcut "multiplying related numbers.")

Calendar

For use with the activities on pages 47 and 48 in the Algebra-based Tricks section.

January

Sun	Mon	Tue	Wed	Thu	Fri	Sat
1	2	3	4	5	6	7
8	9	10	11	12	13	14
15	16	17	18	19	20	21
22	23	24	25	26	27	28
29	30	31				

February

Sun	Mon	Tue	Wed	Thu	Fri	Sat
			1	2	3	4
5	6	7	8	9	10	11
12	13	14	15	16	17	18
19	20	21	22	23	24	25
26	27	28				

March

Sun	Mon	Tue	Wed	Thu	Fri	Sat
			1	2	3	4
5	6	7	8	9	10	11
12	13	14	15	16	17	18
19	20	21	22	23	24	25
26	27	28	29	30	31	

April

Sun	Mon	Tue	Wed	Thu	Fri	Sat
						1
2	3	4	5	6	7	8
9	10	11	12	13	14	15
16	17	18	19	20	21	22
23	24	25	26	27	28	29
30						

May

Sun	Mon	Tue	Wed	Thu	Fri	Sat
	1	2	3	4	5	6
7	8	9	10	11	12	13
14	15	16	17	18	19	20
21	22	23	24	25	26	27
28	29	30	31			

June

Sun	Mon	Tue	Wed	Thu	Fri	Sat
				1	2	3
4	5	6	7	8	9	10
11	12	13	14	15	16	17
18	19	20	21	22	23	24
25	26	27	28	29	30	

July

Sun	Mon	Tue	Wed	Thu	Fri	Sat
						1
2	3	4	5	6	7	8
9	10	11	12	13	14	15
16	17	18	19	20	21	22
23	24	25	26	27	28	29
30	31					

August

Sun	Mon	Tue	Wed	Thu	Fri	Sat
		1	2	3	4	5
6	7	8	9	10	11	12
13	14	15	16	17	18	19
20	21	22	23	24	25	26
27	28	29	30	31		

September

Sun	Mon	Tue	Wed	Thu	Fri	Sat
					1	2
3	4	5	6	7	8	9
10	11	12	13	14	15	16
17	18	19	20	21	22	23
24	25	26	27	28	29	30

October

Sun	Mon	Tue	Wed	Thu	Fri	Sat
1	2	3	4	5	6	7
8	9	10	11	12	13	14
15	16	17	18	19	20	21
22	23	24	25	26	27	28
29	30	31				

November

Sun	Mon	Tue	Wed	Thu	Fri	Sat
			1	2	3	4
5	6	7	8	9	10	11
12	13	14	15	16	17	18
19	20	21	22	23	24	25
26	27	28	29	30		

December

Sun	Mon	Tue	Wed	Thu	Fri	Sat
					1	2
3	4	5	6	7	8	9
10	11	12	13	14	15	16
17	18	19	20	21	22	23
24	25	26	27	28	29	30
31						

Calculation Shortcuts I

Consider two numbers like 43 and 37, which are both 3 away from 40. The following shortcut may be used to make the multiplication of these two numbers easier.

Multiplying Related Numbers

	Example	Working
☛ ◆ Choose two numbers to multiply that are both the same amount away from a multiple of 10, one higher and one lower.	43×37	
◆ Show their relationship to the multiple of 10.	$(40+3)(40-3)$	
◆ Multiply the numbers to give:	$40^2 - 3^2$	
◆ It is now easy to mentally square these numbers.	$1{,}600 - 9$	
◆ This calculation will give you the answer.	$1{,}600 - 9 = 1{,}591$	

Try • Try these problems using the shortcut method.

(a) $33 \times 27 =$ (d) $68 \times 72 =$

(b) $44 \times 36 =$ (e) $398 \times 402 =$

(c) $51 \times 49 =$ (f) $797 \times 803 =$

Check your answers with your calculator.

Calculation Shortcuts II

Squaring or multiplying a two-digit number by itself may be made easier using the following shortcut. $24^2 = 24 \times 24 = ?$

Squaring any two-digit number

	Example	Working
☛ ◆ Choose a two-digit number.	24	
◆ Work out what number you would need to add or subtract to get to the nearest multiple of 10.	4 (difference)	
◆ Write a new easier sum for your square number following these steps:		
• Start with the nearest multiple of 10.	20	
• Multiply by your original number plus the difference.	20×28	
• Add the difference squared to this equation.	$(20 \times 28) + 4^2$	
◆ Simplify the equation.	$560 + 16$	
◆ This sum will give you the answer to your square number.	$560 + 16 = 576$	
	$24^2 = 576$	

Try • Try these problems using the shortcut method.

(a) 23^2 (d) 41^2 (g) 52^2

(b) 32^2 (e) 43^2 (h) 67^2

(c) 35^2 (f) 49^2 (i) 94^2

Check your answers with your calculator.

Square

	Example	Working
☛◆ Think of a number.	8	
◆ Add the number to itself.	8 + 8 = 16	
◆ Subtract it from itself.	8 − 8 = 0	
◆ Multiply it by itself.	8 × 8 = 64	
◆ Divide it by itself.	8 ÷ 8 = 1	
◆ Add your four answers together.	16 + 0 + 64 + 1 = 81	

(Note: $81 = 9^2$ or $(8 + 1)^2$)

◆ Repeat the procedure using different starting numbers.

Try • Try some more numbers to see if this is always the case.

☛ What do you notice about the results?

A Trick for Squares

	Example	Working
☛◆ Choose a number.	23	
◆ Square it.	23 × 23 = 529	
◆ Add 1 to your original number.	23 + 1 = 24	
◆ Square the sum.	576	
◆ Subtract the smaller number from the larger.	576 − 529 = 47	
◆ Subtract 1.	47 − 1 = 46	
◆ Divide the resulting number by 2.	46 ÷ 2 = ?	

◆ Repeat the steps above for several starting numbers.
(You may need a calculator to help you.)

☛ What always happens? What do you notice?

Didax Educational Resources ~ www.didax.com

Digit Shuffle

	Example	Working
☛ ♦ Choose a two-digit number, where each digit is different.	27	
♦ Write down all the numbers that may be formed by changing the positions of the digits.	27, 72	
♦ Add them.	27 + 72 = 99	
♦ Find the sum of the digits in the original number.	2 + 7 = 9	
♦ Divide the total by the sum of the original digits.	99 ÷ 9 = ?	
♦ Repeat the steps above using different two-digit numbers.		
♦ Do you always get the same result? YES/NO		

Try
- Try some three-digit numbers, where each digit is different.
- What do you think the result will be for a four-digit number? Test it!
- Try numbers where some of the digits are the same.

☛ What do you notice?

Finding the Lost Digit

	Example	Working
☛ ♦ Write down any three-digit number.	623	
♦ Cross out one digit (not a 0) to leave a two-digit number.	6̷2̷3 63	
♦ Subtract the two-digit number from the three-digit number.	623 – 63 = 560	
♦ Add up the digits of the remaining number until a single-digit number remains.	5 + 6 + 0 = 11 1 + 1 = 2	
♦ Repeat the above steps using other three-digit numbers.		

☛ What do you notice?

Triple-digit Division

	Example	Working
◆ Choose a three-digit number where all three digits are the same.	555	
◆ Add the digits.	5 + 5 + 5 = 15	
◆ Divide your original number by your answer.	555 ÷ 15 = ?	
◆ Record your answer.		
◆ Try again using different three-digit numbers where all three digits are the same.		

☛ What do you notice?

Try Do you think this will occur with all three-digit numbers where all the digits are the same? YES/NO Try it.

111 ☐   222 ☐   333 ☐
444 ☐   555 ☑   666 ☐
777 ☐   888 ☐   999 ☐

Seeing Double

	Example	Working
☛◆ Write down any three single-digit numbers.	4, 6, 1	
◆ Use these three numbers to make six two-digit numbers.	46, 41, 61, 64, 14, 16	
◆ Add the six two-digit numbers.	242	
◆ Add the original three numbers.	4 + 6 + 1 = 11	
◆ Divide the larger number by the smaller.	242 ÷ 11 = ?	
◆ Write down your answer.		
◆ Repeat the steps above using three different single-digit numbers.		

Try • Try using consecutive single-digit numbers; e.g. 1, 2, 3 or 2, 3, 4.

☛ What do you notice? Try to explain why it happens.

Magic with Math ~ 43

Do or Die

	Example	Working
☛♦ Throw two dice and note the numbers.	3, 5	
♦ Multiply these two numbers.	3 x 5 = 15	
♦ Lift the dice and note the two numbers on the bottom of the dice.	4, 2	
♦ Multiply these two numbers.	4 x 2 = 8	
♦ Multiply the top number of the first die by the bottom number of the second die.	3 x 2 = 6	
♦ Multiply the top number of the second die by the bottom number of the first die.	5 x 4 = 20	
♦ Add your four results.	15 + 8 + 6 + 20 = ?	
♦ Roll the dice again and repeat the procedure.		
♦ Try a few more times and discuss the results with your partner.		

☛ What do you notice?

Dice Dropping

	Example	Working
☛♦ Roll three dice.		
♦ Enter the three-digit number produced into a calculator. The order of the numbers is not important.	624	
♦ Enter the numbers shown on the bottom of each die in the same order to make a six-digit number.	624,153	
♦ Divide the six-digit number by 37.	624,153 ÷ 37 = 16,869	
♦ Divide the result by 3.	16,869 ÷ 3 = 5,623	
♦ Subtract 7.	5,623 – 7 = 5,616	
♦ Divide by 9.	5,616 ÷ 9 = ?	
♦ Repeat the steps above and note what happens.		

☛ What do you notice about the result? Try to explain why this happens.

Double Trouble

	Example	Working
♦ Write down any three different single-digit numbers.	3, 4, 7	
♦ Use these three numbers to make nine two-digit numbers. Digits may be repeated.	33, 44, 77, 34, 43, 73, 37, 47, 74	
♦ Add the nine two-digit numbers.	33 + 44 + 77 + 34 + 43 + 73 + 37 + 47 + 74 = 462	
♦ Add the original numbers.	3 + 4 + 7 = 14	
♦ Divide the larger number by the smaller.	462 ÷ 14 = ?	
♦ Write down your answer.		
♦ Repeat the steps above using three different single-digit numbers.		

Try
- Try starting with three odd numbers, then three even numbers.
- Try using consecutive single-digit numbers: (1, 2, 3), (2, 3, 4), (3, 4, 5) and so on.

☛ What do you notice? Try to explain why it happens.

3,087

	Example	Working
♦ Write a four-digit number where each digit is one less than the previous digit.	7,654	
♦ Reverse this number.	4,567	
♦ Subtract the smaller number from the larger.	7,654 − 4,567	
♦ Record the answer.		
♦ Repeat the steps above with other four-digit numbers where each digit is one less than the previous digit.		

Try
- Try the same procedure using three-digit and five-digit numbers.

☛ Write about what you notice.

Cycling Digits

	Example	Working
☛♦ Write down a four-digit number.	9,351	
♦ Cycle the digits by moving the digit in the thousands place to the hundreds place. The digit in the hundreds place is then moved to the tens place. The tens digit is moved to the ones place. The ones digit is then moved to the thousands place.	1,935	
♦ Continue cycling the digits until four numbers have been produced.	5,193 3,519	
♦ Add the four numbers.	9,351 + 1,935 + 5,193 + 3,519 = 19,998	
♦ Find the sum of the original four digits.	9 + 3 + 5 + 1 = 18	
♦ Divide the total from adding the four numbers by the sum of the four digits.	19,998 ÷ 18 = ?	
♦ Record the result.		
☛♦ Repeat the steps above with other four-digit numbers.		

- Try the above for three-digit numbers or five-digit numbers.
- What happens if you begin with a number where all four digits are the same?

☛ Write about what you notice.

Mathemagic

	Example	Working
☛♦ Choose a two-digit number where both digits are the same.	77	
♦ Add the digits.	7 + 7 = 14	
♦ Divide the original number by the sum of its digits.	77 ÷ 14 = ?	
♦ Repeat the steps with other numbers where both digits are the same.		

- Try starting with a three- and four-digit number.
- What happens with nine-digit numbers?

☛ What do you notice about the result?

46 ~ Magic with Math

Calendar Magic

▶◆ Choose any month from the year and draw a 3x3 box around any nine dates.

	January					
sun	mon	tue	wed	thu	fri	sat
	1	2	3	④	5	6
7	8	9	10	11	12	13
14	15	16	17	18	19	20
21	22	23	24	25	26	27
28	29	30	31			

Example   Working

- ◆ Choose the smallest number in the box.   4
- ◆ Add 8 to it.   4 + 8 = 12
- ◆ Multiply the sum by 9.   9 × 12 = ?
- ◆ Record your answer.
- ◆ Add the nine numbers in the box.

  4 + 5 + 6 + 11 + 12 + 13 + 18 + 19 + 20 = ?

▶ How does your answer compare to the sum of the numbers in the box?

Try
- Try other months.
- Does it work for every month in the year? YES/NO

Magic Months

▶◆ Work with a partner. Keep this sheet hidden from him/her while you work your "magic" trick.

	January					
sun	mon	tue	wed	thu	fri	sat
	1	2	3	4	5	6
7	8	9	10	11	12	13
14	15	16	17	18	19	20
21	22	23	24	25	26	27
28	29	30	31			

Example   Working

- ◆ Ask your partner to secretly choose a 3x3 block from any month on the calendar, and follow this procedure:
  - Add the nine numbers:

    11 + 12 + 13 + 18 + 19 + 20 + 25 + 26 + 27 = 171

  - Write the total and show it to you.
- ◆ Using your "Secret Solution," offer to "read" his/her mind to find out the 3x3 block in the chosen month.

Secret Solution

Mentally divide the total by 9, which will reveal the middle number in the block.
Subtract 8 to determine the first number in the block.
The rest may be worked out using the sequence of the calendar.

Didax Educational Resources ~ www.didax.com   Magic with Math ~ 47

Cool Calendar

☛ ◆ Choose any month from the calendar.
   ◆ Find four dates that form a 2x2 block and draw a box around them.

			January			
sun	mon	tue	wed	thu	fri	sat
	1	2	3	4	5	6
7	8	9	10	11	12	13
14	15	16	17	18	19	20
21	22	23	24	25	26	27
28	29	30	31			

	Example	Working
◆ Add the four numbers.	18 + 19 + 25 + 26 = 88	
◆ Divide the sum by 4.	88 ÷ 4 = 22	
◆ Subtract 4 from this result.	22 − 4 = ?	
◆ Discuss what you notice with a partner.		
◆ Repeat for other dates and months.		

☛ Write about what you notice.

Your Puzzle

☛ The calendar has a uniform structure. This helps when writing your own puzzle. You may like to try writing a puzzle that involves a row (7 numbers) or a column (5 numbers) from the calendar.

48 ~ **Magic with Math**   www.didax.com ~ Didax Educational Resources

Explanations

Calculation Shortcuts I
– page 40

This shortcut is based on the difference of squares;
i.e. $(a - b)(a + b) = a^2 - b^2$.

(a) 33×27 $= (30 + 3)(30 - 3)$
                   $= 900 - 9$
                     $= 891$

(b) 44×36 $= (40 + 4)(40 - 4)$
                   $= 1{,}600 - 16$
                   $= 1{,}584$

(c) 51×49 $= (50 + 1)(50 - 1)$
                   $= 2{,}500 - 1$
                   $= 2{,}499$

(d) 68×72 $= (70 - 2)(70 + 2)$
                   $= 4{,}900 - 4$
                   $= 4{,}896$

(e) 398×402 $= (400 - 2)(400 + 2)$
                     $= 160{,}000 - 4$
                   $= 159{,}996$

(f) 797×803 $= (800 - 3)(800 + 3)$
                     $= 640{,}000 - 9$
                   $= 639{,}991$

Calculation Shortcuts II
– page 40

(a) $23^2 = (20 \times 26) + 3^2$
       $= 520 + 9$
       $= 529$

(b) $32^2 = (30 \times 34) + 2^2$
       $= 1{,}020 + 4$
       $= 1{,}024$

(c) $35^2 = (30 \times 34) + 5^2$
       $= 1{,}200 + 25$
       $= 1{,}225$

(d) $41^2 = (40 \times 42) + 1^2$
       $= 1{,}680 + 1$
       $= 1{,}681$

(e) $43^2 = (40 \times 46) + 3^2$
       $= 1{,}840 + 9$
       $= 1{,}849$

(f) $49^2 = (50 \times 48) + 1^2$
       $= 2{,}400 + 1$
       $= 2{,}401$

(g) $52^2 = (50 \times 54) + 2^2$
       $= 2{,}700 + 4$
       $= 2{,}704$

(h) $67^2 = (70 \times 64) + 3^2$
       $= 4{,}480 + 9$
       $= 4{,}489$

(i) $94^2 = (90 \times 98) + 4^2$
      $= 8{,}820 + 16$
      $= 8{,}836$

Square – page 41

The steps in this puzzle essentially describe the binomial expansion. For example: $(8 + 1)^2$
$= (8 + 1)(8 + 1)$
$= (8 \times 8) + (8 \times 1) + (1 \times 8) + (1 \times 1)$
$= 64 + 8 + 8 + 1$

Note how the steps in the expansion relate to the instructions given. Adding the number to itself in step 1 covers the middle of the expansion (in this case 8 + 8). Step 2 is a red herring–the result of zero has no bearing on the outcome. Multiplying by itself covers the first part of the expansion and dividing by itself covers the last part. All that is left to do is add all the parts.

A Trick for Squares
– page 41

This "trick" is simply another application of the binomial expansion.

Square the original number: 23^2
Add 1 and square: $(23 + 1)(23 + 1) = 23^2 + 2(23) + 1$
Find the difference: $23^2 + 2(23) + 1 - 23^2 = 2(23) + 1$
Subtract 1: $2(23) + 1 - 1 = 2(23)$
Halve the number: $2(23) \div 2 = 23$

It always returns you to the starting number.

Digit Shuffle – page 42

When following this procedure with a two-digit number the answer is always 11.
The result for three-digit numbers is always 222.
The result for four-digit numbers is always 6,666.
Let the two digits be a and b.
The numbers formed according to the instructions:
$10a + b$ and $10b + a$.
Adding them we get: $11a + 11b$ or $11(a + b)$
                     or 11(sum of the original digits)

Therefore, when the total is divided by this sum the answer will always be 11. Similar reasoning may be applied to three-digit numbers. If the three digits are represented by a, b and c the following combinations are formed by the instruction to write down all the numbers that may be formed by changing the positions of the digits.

abc	e.g. 123
$100a + 10b + c$	123
$100a + 10c + b$	132
$100b + 10a + c$	213
$100b + 10c + a$	231
$100c + 10b + a$	321
$100c + 10a + b$	312

Adding the above we get $222a + 222b + 222c$ or $222(a + b + c)$. Dividing by the sum of the original digits $(a + b + c)$ leaves 222. The same reasoning may be applied to the four-digit numbers.

Explanations

Finding the Lost Digit
– page 42

The final number is the crossed out digit.
Although the following is not a full explanation, it outlines the basic principle behind the trick.

digit sum of $100a + 10b + c = a + b + c$
digit sum of $10a + c \quad\quad = a + c$
digit sum of difference $\quad = b$

Triple-digit Division – page 43

The answer is always 37.
Let the digit be represented by a.
Therefore, a three-digit number would be $100a + 10a + a$ or $111a$.
Adding the digits gives $a + a + a$ or $3a$
$111a \div 3a = 37$ always leaving 37 as your answer.

Seeing Double – page 43

Let a, b and c represent the single-digit numbers.
The six two-digit numbers are formed thus:

$\quad\quad 10a + b$
$\quad\quad 10a + c$
$\quad\quad 10b + a$
$\quad\quad 10b + c$
$\quad\quad 10c + a$
$\quad\quad 10c + b$

Adding the terms gives: $22a + 22b + 22c$ or $22(a + b + c)$
Dividing by $(a + b + c)$ gives 22 every time.

Do or Die – page 44

The answer is always 49. A little algebra helps to show why.

If we call the numbers on the top of each die a and b, then the numbers on the bottom of each die will be $7 - a$ and $7 - b$, because the numbers on opposite faces of a die always add up to 7.
Multiplying the top numbers: $a \times b = ab$.
Multiplying the bottom numbers:
$(7 - a) \times (7 - b) = 49 - 7a - 7b + ab$
(A knowledge of how to multiply binomials is required to understand this part. This trick could be used to develop the topic of multiplying binomials.)
Multiplying the top number from the first die and bottom number from the second die:
$a \times (7 - b) = 7a - ab$.
Multiplying the top number from the second die and the bottom number from the first die:
$b \times (7 - a) = 7b - ab$.
Combining the results we have: $(ab) + (49 - 7a - 7b + ab) + (7a - ab) + (7b - ab)$,
which, when simplified, leaves 49.

Dice Dropping – page 44

If the value on the top of the dice are denoted by a, b and c, then the number produced by entering the three numbers shown on top of the dice and the three on the bottom would be represented by:
$100,000a + 10,000b + 1,000c + 100(7 - a) + 10(7 - b) - (7 - c)$ which, when simplified, gives
$99,900a - 9,990b + 999c + 777$.
Removing a common factor of 111 gives:
$111(900a + 90b + 9c + 7)$.
Dividing by 37 and then by 3 is the same as dividing by 111, which leaves: $900a + 90b + 9c + 7$;
subtracting 7 leaves: $900a + 90b + 9c$;
and dividing by 9 gives $100a + 10b + c$, the original three numbers.

Double Trouble – page 45

This mathematical novelty is similar to **Seeing Double**. Let a, b and c represent the single-digit numbers. The 9 two-digit numbers formed are:

$10a + a \quad\quad 10a + b \quad\quad 10a + c$
$10b + a \quad\quad 10b + b \quad\quad 10b + c$
$10c + a \quad\quad 10c + b \quad\quad 10c + c$

Adding the terms we get
$33a + 33b + 33c = 33(a + b + c)$.
Dividing by $(a + b + c)$, the original three digits equals 33 every time.

3,087 – page 45

All four-digit numbers produce 3,087. Three-digit numbers produce 198, five-digit numbers produce 15,873. A little algebra shows why four-digit numbers produce 3,087.
Let a represent the digit in the thousands place, then the first instruction gives:

$\quad 1,000a + 100(a - 1) + 10(a - 2) + (a - 3)$
$\quad - 1,000(a - 3) + 100(a - 2) + 10(a - 1) + a$
$\quad$ (second and third instructions)
$\quad = 3,087$

Explanations

Cycling Digits – page 46

If we let the four digits be represented by a, b, c and d, then cycling the digits gives:

$(1{,}000a + 100b + 10c + d)$
$+ (1{,}000d + 100a + 10b + c)$
$+ (1{,}000c + 100d + 10a + b)$
$+ (1{,}000b + 100c + 10d + a)$
$= (1{,}111a + 1{,}111b + 1{,}111c + 1{,}111d)$
or $1{,}111(a + b + c + d)$.

Dividing by the sum of all the digits $(a + b + c + d)$ leaves $1{,}111$.

If you begin with a three-digit number, the result will always be 111, and if you begin with a five-digit number, the result will always be 11,111.

$111{,}111 \div 37 = 3{,}003$ etc.

Starting with four digits that are the same (or any number of digits) will not make any difference.

Mathemagic – page 46

For two-digit numbers, the result is always 5.5. For three-digit numbers the result is always 37. The result for four-digit numbers is always 277.75. A nine-digit number produces a result of 12,345,679. A little algebra shows why this works.

Choose a two-digit number where both digits are the same: $(10a + a) = 11a$
Add the digits: $2a$
Divide the original by the sum of its digits:
$11a \div 2a = 5.5$

Calendar Magic – page 47

Adding 8 to the smallest number in the box and multiplying by 9 gives the same result as adding the nine dates in the box. This procedure will work for any month and any year.

We can see why it works if we use d to represent the smallest date in the set of 9 dates:

d	d + 1	d + 2
d + 7	d + 8	d + 9
d + 14	d + 15	d + 16

Adding 8 to the smallest number gives $d + 8$
Multiplying by 9 gives $9(d + 8)$ or $9d + 72$
Adding the nine dates gives:
$d + (d + 1) + (d + 2) + (d + 7) + (d + 8) + (d + 9) +$
$(d + 14) + (d + 15) + (d + 16)$ ... or $9d + 72$

Magic Months – page 47

If the 3x3 block of dates is represented by:

d	d + 1	d + 2
d + 7	d + 8	d + 9
d + 14	d + 15	d + 16

Adding the nine dates:
$d + (d + 1) + (d + 2) + (d + 7) + (d + 8) + (d + 9) + (d + 14) + (d + 15) + (d + 16)$ gives $9d + 72$ or $9(d + 8)$. Dividing by 9 will give $d + 8$ or the middle date. To work out the surrounding dates, simply add or subtract the appropriate numbers.

Cool Calendar – page 48

Consider the four dates:

d	d + 1
d + 7	d + 8

Add the four dates $d + (d + 1) + (d + 7) + (d + 8)$
                                 $4d + 16$
Divide by 4 $d + 4$
Subtract 4 d (the date at the top left)

Your Puzzle – page 48

Teacher check

Notes

Card Tricks

The card tricks that follow are all self-working. They do not rely on sleight of hand, specially marked decks, or stacking the deck. As long as you follow the sequence of steps outlined in the trick the desired outcome will be achieved.

Dazzle your class with the following trick:

With a standard shuffled deck of cards, ask a student in the audience to name two cards. (Don't worry about naming the suits.)

Now state that you will try to move the two cards together without touching the deck. You might also like to explain that as you are a bit rusty the two cards might be separated by one card.

Start dealing the cards face up until the chosen cards come up.

Note: Now and again this self-working card trick doesn't work. Don't panic. Simply state that you must have been thinking of the wrong pair of cards and restart the trick.

The probability is such that there is an extremely high chance of the cards being next to each other occurring naturally.

Crazy Cards

- ☛♦ *Work with a partner. Keep this sheet hidden from him/her while you work your "magic" trick.*
- ♦ *Deal 15 cards, face up, from an ordinary deck of playing cards, so that three rows of five are formed. For example:*

Note: It is a good practice to deal the cards so that each successive card in the row slightly overlaps the previous card. This makes it easy to maintain the order of the cards.

- ♦ *Ask your partner to choose a card, but to keep it secret. (You may wish to turn your back while your partner shows the rest of the class.) Now follow the steps below:*

  1. Ask your partner to indicate which row contains the chosen card.
  2. Collapse each row, ensuring the order of the cards is maintained. Place the row containing the chosen card between the two remaining rows.
  3. Deal the cards again, dealing them into columns beginning at the top left-hand corner and finishing at the bottom right corner.

1	4	7	10	13
2	5	8	11	14
3	6	9	12	15

  4. Ask your partner to indicate which row now contains the secret card.
  5. Repeat steps 2, 3 and 4 and then collapse the cards again as in step 2.

- ♦ *Announce that you will now reveal the secret card using your "Secret Solution."*

Secret Solution

Count (silently) seven cards from the top of your pack.
The secret card will be the eighth card.

- Try using 21 cards in 3 rows of 7. You will need to deal the cards three times and then count out 11 cards at the end. The secret card will be the twelfth card.

☛ *Discuss with your partner how you think this trick works. Try to explain how it happens below. You may like to mark a card and watch what happens to the position of this card as the trick progresses.*

A Math Challenge is Fun

	Example	Working

♦ Cut a deck of cards close to the middle.
- ♦ Count the cards in the top part of the cut.   26
- ♦ Add the digits.   2 + 6 = 8
- ♦ Count this number of cards up from the bottom of the top part of the deck to reveal the magic card.
- ♦ Now put the deck back together in its original order.
- ♦ The magic card may be found by spelling out the phrase
  A MATH CHALLENGE IS FUN
  and dealing out the cards one at a time as each letter is spoken.
  The magic card will be the final card (L).

Try • Try this trick on someone else. Think of a new phrase to use (count the letters carefully!)

 Write about what you notice and try to explain how this trick works.

Card Shark

- ♦ Work with a partner. Keep this sheet hidden from him/her while you work your "magic" trick.
- ♦ Remove 26 cards from a standard deck.
  - ♦ Split the 26 cards into two piles: one containing 10 cards, the other 16.
  - ♦ Ask a friend to choose a card from the pile of 16 cards, then place his/her card on top of the pile of 10 cards.
  - ♦ Place the remaining 15 cards on top of the pile of 11.
  - ♦ To reveal the chosen card use the "Secret Solution."

Secret Solution

- ♦ Deal the pile of 26 cards alternately into two rows face down.

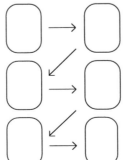

- ♦ Once the 26 cards have been dealt, discard the 13 cards in the left row.
- ♦ Collapse the right-hand row, being careful to maintain the order of the cards.
- ♦ Repeat the above steps until there is only one card left. This will be the chosen card.

Card Caper

The following is a self-working mathematical card trick, which means if you follow the instructions carefully it will work every time.

- ♦ Work with a partner. Keep this sheet hidden from him/her while you work your "magic" trick.
- ♦ Check that you have a complete deck of cards (52) with the jokers removed before you start.
- ♦ Turn over the top card from the pack and place it face up on the table.

- ♦ Look at the value shown on the card, e.g. 8. (If the card is an Ace it has a value of 1, a Jack, 11, a Queen, 12 and a King, 13.) Count from the value shown on the card up to 13. Add a card to the pile each time you say a number. For example, if an 8 was showing you would count, 9, 10, 11, 12, 13, placing five cards in all on top of the 8. Should a King be turned up, no cards would be added.

- ♦ Once you have counted to 13, turn over the pile (face down so the original card—e.g. 8—is on top) and continue the process.
- ♦ Eventually, you will have several piles on the table and you may find that you do not have enough cards to complete a final pile. Simply place these "extra" cards to one side.

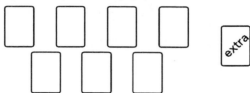

- ♦ Next ask your partner to point to any three piles on the table. Remove all the other piles and add them to the "extra" cards that you placed to one side.

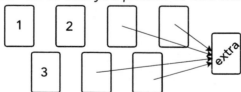

- ♦ Next ask your partner to turn over the top card from any of two of the three piles left on the desk.

- ♦ Note the values of the top two cards, e.g. 7, and Q (12), and combine them, e.g. (12 + 7) 19. Add a further 10 to this number, e.g. 19 + 10 = 29.
- ♦ Count this many cards from the pile of "extra" cards that were placed to one side and hand these to your partner.

- ♦ Finally reveal the value of the top card on the remaining pile by counting the number of cards that are remaining in your hand (the "extra" pile). For example, if 11 cards remain, the top card will be a Jack.

Explanations

Crazy Cards – page 54

Dealing the cards in the sequence outlined ensures that the chosen card always ends up in the center of the middle row; the chosen card is forced to the center by the order in which the cards are dealt. Being in the middle set of five, after the first deal the card is one of those marked.

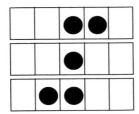

After the second deal, the card is one of:

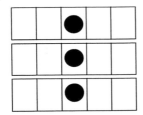

When placed in the middle row, the chosen card is the middle one (i.e. the eighth card).

A Math Challenge is Fun – page 55

This trick only works when the top cut from the deck of cards contains between 20 and 29 cards. The difference between the top number of cards in this part of the deck and the sum of the digits of this top set number will always be 18. The phrase A MATH CHALLENGE IS FUN contains 19 letters and therefore, will always bring you back to the magic card. A little algebra explains why this works.

Let a represent the tens digit.
Let b represent the ones digit.
The two-digit number would be $10a + b$.
The sum of the digits would be $a + b$.
The difference is $9a$.
When the number is in the 20s $a = 2$, and the difference is therefore 18. The 19 letter phrase reveals the actual card.

Variations

Alter the phrase to be spelled out. Any 18- or 19-letter phrase will do. If an 18-letter phrase such as ALL MATH IS WONDERFUL is used, simply count 18 cards and turn the nineteenth card to reveal the magic card.

If the number of cards contained in the cut from the top of the deck contains between 30 and 39 cards then the difference between the top number of cards in the top cut of the deck and the sum of the digits of this top set number will always be 27 (i.e. 9 x 3). Likewise, if the number of cards contained in the top cut from the deck is between 10 and 19, then you will only need to count 9 cards (i.e. 9 x 1) to reveal the magic number.

Card Shark – page 55

Drawing a table which locates the required card in the 16th position on the first sorting will help to show that successive sortings based on removing the left-hand column of cards eventually leave only one card remaining in the right-hand column. The first deal produces 13 cards in each row. The left row is discarded. The second deal produces seven cards in the left row and six in the right. The third produces three cards in each row and so on until a single card is left.

Card Caper – page 56

If the card showing is represented by a, then the number of cards in the pile will be $14 - a$. For example, if an 8 is turned over, five more cards will be added, making six in the pile. After the first pile is produced, there should be $52 - (14 - a)$ cards or $38 + a$ cards left. Similarly, the second pile will contain $14 - b$ cards, leaving $24 + a + b$ cards, the third will contain $14 - c$ cards, leaving $10 + a + b + c$ cards, the fourth pile will contain $14 - d$ cards, leaving $-4 + a + b + c + d$ cards, a fifth pile would contain $14 - e$ cards, leaving $-18 + a + b + c + d + e$ cards.

Imagine your partner chooses the following piles: $14 - a$, $14 - c$ and $14 - d$.

The top cards from two of these piles might be a and d.

Adding 10 to these produces $a + d + 10$.

Collecting the remaining piles and adding the leftover cards produces $(14 - b) + (14 - e) + (-18 + a + b + c + d + e) = 10 + a + c + d$.

Counting out $a + d + 10$ will leave c cards, which is the value of the card on top of the remaining pile.

Notes

A Mixed Bag

The tricks in this section make use of a variety of mathematics. Some of the topics covered include topology; area and a knowledge of dice.

The tricks in this section could easily be used to introduce a topic such as area. For example, the **Missing Area** tricks may be used as a means of encouraging students to work out the area of rectangles and triangles and then be used to introduce the formula for determining the area of a trapezoid.

The **Thrice Dice** trick relies on the fact that the opposite faces on a fair die add to seven. This trick may be used to stimulate the investigation of the nets that may be folded to form a cube. Once all the nets have been found the children could be challenged to label the net with the numbers 1 to 6, so that when folded opposite faces add to seven.

A third group of tricks found in this section, **Stepping Through Paper**, **Tied up in Knots**, and those involving the möbius strip, are related to topology – a branch of geometry. Students come across topology when working with networks. (A knowledge of networks is required when producing electronic circuitry or setting up computer networks.) Many puzzle books contain problems like "Is it possible to trace over this diagram without lifting your pen?"

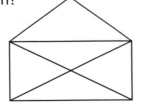 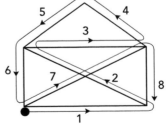

Other aspects of topology include the study of mazes, also popular in puzzle books and colouring regions, such as those found on a map. The möbius strip is of particular interest because it effectively only has one side. Some belt systems and conveyor belts incorporate möbius bands as this ensures even wear on both sides of the belt.

Didax Educational Resources ~ www.didax.com   Magic with Math ~ 59

Grids

For use with the activities on pages 63 and 64 in the Mixed Bag section.

Stepping Through Paper

- ◆ You will need a sheet of 8½" x 11" paper and a pair of scissors.
- ◆ Fold the paper in half lengthwise.

- ◆ Cut from the folded edge toward the outside edge of the paper. Make sure the cuts are at least 2 cm apart, stopping short of the edge.

- ◆ Next, cut back from the outside edge towards the fold. Make sure you cut between your first set of cuts and that you stop short of the fold.

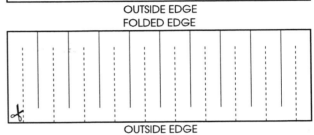

- ◆ Use the scissors to cut along the fold in the paper between the two marked points.

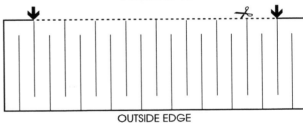

- ◆ Open the paper up and step through it.

Tied up in Knots

To perform the following trick you will need two pieces of string each about one yard long.
- ◆ Find a partner.
- ◆ Loosely tie a piece of string around each of your partner's wrists.
- ◆ Tie a second piece of string around one of your wrists. Loop the string around the piece tied around your partner's wrists and tie the string to your other wrist. (See the diagram below.)
- ◆ Try to separate yourselves without cutting or untying the string.

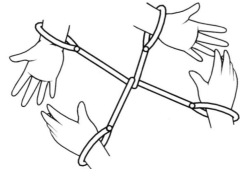

A Shape with a Twist I

Möbius strips were first invented by German astronomer and mathematician August Ferdinand Möbius (1790 – 1868).

♦ To make your own möbius strip, follow the instructions.
  • Cut the strip of paper from the right side of this worksheet.

  • Form a half twist in the strip by turning one end over (turn one end 180°).

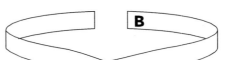

  • Join the ends together with tape to form a shape. This is called a möbius strip.

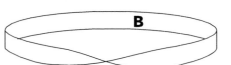

♦ Starting at A, draw a line on the band until your reach B and continue the line until you return to A.
♦ Did the line need to cross an edge? YES/NO
♦ Does the möbius strip have one or two sides?

• Try creating a möbius strip using a full twist.

A Shape with a Twist II

♦ Cut a strip of paper 30 cm by 2 cm (or use the strip from A Shape with a Twist I) and mark a line down the center.
♦ Give your strip a half twist and join the ends together with tape to form a möbius strip.
♦ Predict what you think will happen when you cut along the center line.

♦ Now, cut along the center line. What happens?

♦ Try marking two lines along the center of a new strip of paper and repeating the above process.
♦ Predict what you think will happen when you cut along the two lines.

♦ Now cut along the two lines. What happens?

62 ~ **Magic with Math**   www.didax.com ~ Didax Educational Resources

Missing Area I

- ◆ Mark out an 8x8 square on a piece of graph paper.
- ◆ What is the area of the square? Area: ☐
- ◆ Cut out the square and reform to make a rectangle.

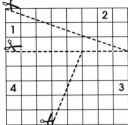

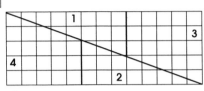

- ◆ Now work out the area of the rectangle. Area: ☐

☛ What do you notice? Try to explain how you think this has happened.

Number Words

	Example	Working
☛ ◆ Choose a number.	54	
◆ Write it in words.	fifty-four	
◆ Count the letters.	9	
◆ Write that number in word form.	nine	
◆ Count the letters.	4	

(There is no need to go beyond 4 because the word "four" contains 4 letters.)

- ◆ Choose some starting numbers of your own and note what happens when you follow the above procedure.

Try Start with the following numbers and see what happens: 44, 72, 77, 23.

☛ Write about what you notice.

Missing Area II

- ♦ Mark out an 8x8 square as shown on a piece of graph paper.
- ♦ Cut out the square and reform to make four pieces as shown.
- ♦ Two different isosceles triangles may be formed using these four pieces. Make each triangle.
- ♦ Work out the area of each triangle by using the formula:
  (Area of a triangle = $\frac{1}{2}$ base x height).
- ♦ Record your results.

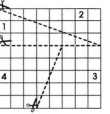

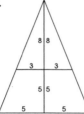 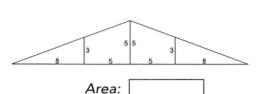

Area: _____   Area: _____

- ♦ Now form a trapezoid with the four pieces and find the area.
  (The area of a trapezoid is found by averaging the lengths of the base and top of the trapezoid, and multiplying by the perpendicular height.) Record your result.

  Area: _____

- ♦ The four pieces may be fitted together to form a parallelogram. Find the area of the parallelogram. Record your result.
  (Area of a parallelogram = base x perpendicular height)

  Area: _____

☛ Write about what you notice. Try to explain why you think this happens.

Thrice Dice

- ♦ Work with a partner. Keep this sheet hidden from him/her while you work your "magic" trick.
- ♦ Turn your back and ask your partner to roll three dice and stack them on top of each other.
- ♦ While your back is turned, ask your partner to add up the numbers of the five hidden faces of the dice and keep the total secret.
- ♦ Look at the stack of dice and note the top face on the stack.
- ♦ Using the "Secret Solution" announce the total to your partner after some show of consternation (and some simple mental addition).

Secret Solution

To determine the total, subtract the number shown on the top face from 21.

Explanations

Stepping Through Paper
– page 61

If you wish to make the effect more dramatic, you can start with a much smaller sheet of paper and still walk through it. There are several ways that a sheet of paper may be cut in order to walk through it.

For example:

Draw lines on your sheet of paper and cut along these lines. Start cutting the outside lines first. Stop cutting at each of the dots.

Note: By cutting at smaller intervals, a larger opening is created.

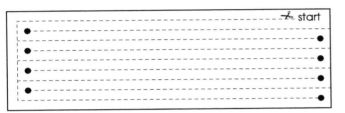

Tied up in Knots – page 61

This trick is based on an understanding of topology.

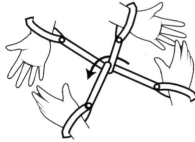

To free the tied hands above, first form a loop with one of the pieces of string. Thread the loop through the string tied around the other person's wrist as per the diagram, keeping the string tight.

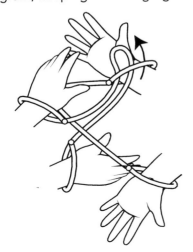

Pull the loop down over the other person's hand.

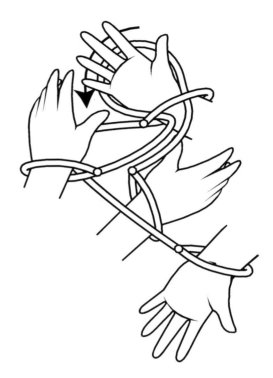

Pull your string tight to disengage.

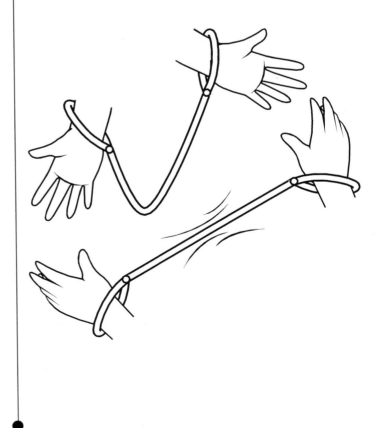

Explanations

A Shape with a Twist I
– page 62

When you form the möbius strip and draw a line from A to B you do not cross any edges. A möbius strip has only one edge.

A Shape with a Twist II
– page 62

When you cut the möbius strip down the center line you form one larger single band.
When you cut along two lines you form two loops.

Missing Area I – page 63

The square has an area of 64 square units, while it would appear the rectangle has an area of 65 square units (13x5). The reason for the apparent discrepancy is that the pieces along the diagonal of the rectangle do not quite fit together. There is a gap in the shape of a parallelogram which has an area of 1 square unit, which makes up the difference.

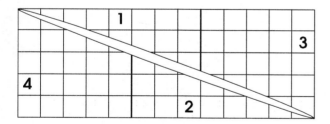

Number Words – page 63

Once you reach 4, an endless loop forms because the word four is made up of 4 letters.

Missing Area II – page 64

The area of both triangles is 65 square units.
Area of Trapezoid
= average of parallel sides x perpendicular height
= [(5 + 11) ÷ 2] x 8
= 64 square units

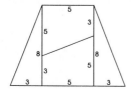

Area of Parallelogram
= base x perpendicular height
= 8 x 8
= 64 square units

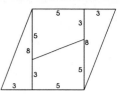

The reason for this apparent contradiction lies along the diagonal.
The pieces in the traingles don't really fit together and so a small parallelogram with an area of 1 square unit is formed. You may also like to use trigonometry to work out the angle measurements where the triangle and the trapezoid join to form a straight line. You will find the two angles a and b add to slightly more than 180°.

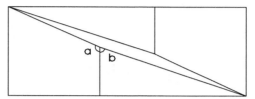

Thrice Dice – page 64

The opposite sides of a die will always add up to 7. Therefore, the three opposite sides of three dice will add up to 21. To determine the answer, simply look at the number on top of the top die and subtract it from 21.

Teacher Tricks

All the tricks in this section require some preparation, either in terms of creating some props (many of which simply need to be copied or enlarged) or memorizing a few steps.

The mathematics behind many of these tricks requires a knowledge of patterns and sequences such as:

- The powers of two, $2^0, 2^1, 2^2, 2^3$...
- The Fibonacci sequence 1, 1, 2, 3, 5, 8, 13, which is formed by adding the two previous terms in the sequence.

A simple trick based on an understanding of odd and even numbers is illustrated below.

You will need five coins.

Ask a student to divide five coins between both hands (obviously there will be an odd number of coins in one hand and an even number in the other hand).

Ask the student to :

- Multiply the number of coins in his/her right hand by 2.
- Multiply the number of coins in his/her left hand by 3.
- Add the two results.
- Tell you the total.

An even answer means the even number of coins are in the students left hand.

An odd answer means the odd number of coins are in the students left hand.

Reveal without looking which hand contains the even number of coins and which the odd.

Mind-reading Trick

- Using the six number pages on pages 71 to 76, you are going to "read" a student's mind.
- Ask a student to choose a number between 1 and 63. This number should be kept secret.
- Show the student each number page in turn and ask whether his/her number appears on that particular page.
- If the student says that the number appears on the page, note the number in the top left-hand corner of the number page. Add these numbers as you go.
- The total is the secret number.
- For example, if the secret number is 23, it appears on the following pages.

①	23	45
3	25	47
5	27	49
7	29	51
9	31	53
11	33	55
13	35	57
15	37	59
17	39	61
19	41	63
21	43	

②	23	46
3	26	47
6	27	50
7	30	51
10	31	54
11	34	55
14	35	58
15	38	59
18	39	62
19	42	63
22	43	

④	23	46
5	28	47
6	29	52
7	30	53
12	31	54
13	36	55
14	37	60
15	38	61
20	39	62
21	44	63
22	45	

⑯	27	54
17	28	55
18	29	56
19	30	57
20	31	58
21	48	59
22	49	60
23	50	61
24	51	62
25	52	63
26	53	

1 + 2 + 4 + 16 = 23

Explanation

The number pages are produced in such a way that the numbers from 1 to 63 may be formed by using various combinations of 1, 2, 4, 8, 16 and 32. These numbers are shown on the top left of each card. As the person indicates which page contains the number, simply note the number on the top left of the page. Add all the numbers as you go. The total will be the person's secret number.

This trick uses base two, similar to computers.

For example: $23 = (1 \times 2^4) + (0 \times 2^3) + (1 \times 2^1) + (1 \times 2^0)$
$= (1 \times 16) + (0 \times 8) + (1 \times 4) + (1 \times 1)$

Computers would record it as 10111. Thus, adding the base two values (top left of each page) gives the value of the chosen number. The pages are made up by combining the powers of two.

Number	2^0	2^1	2^2	2^3	2^4	2^5
	1	2	4	8	16	32
1	✓					
2		✓				
3	✓	✓				
4			✓			
5	✓		✓			
6		✓	✓			
7	✓	✓	✓			
8				✓		
9	✓			✓		
10		✓		✓		
11	✓	✓		✓		
12			✓	✓		
13	✓		✓	✓		
14		✓	✓	✓		

Two appears on only one page – the one with 2 in the top left corner.

Seven appears on three pages – the ones with 1, 2 and 4 in the top left corner.

Fourteen appears on three pages – the ones with 2, 4 and 8 in the top left corner.

Making Mind-reading Pages

➡◆ To work out which numbers to place on a simple set of five mind-reading number pages, complete the following table.

	2^0	2^1	2^2	2^3	2^4
	1	2	4	8	16
1					
2					
3					
4					
5					
6					
7	✓	✓	✓		
8					
9					
10					
11					
12					
13					
14					
15					
16					✓
17					
18					
19					
20					
21					
22					
23					
24					
25					
26					
27					
28					
29					
30					
31	✓	✓	✓	✓	✓

⬅ *Seven should appear on three pages – the ones with 1, 2 and 4, because 1 + 2 + 4 = 7*

⬅ *Sixteen should appear on one page – the one with 16 showing in the top left corner.*

⬅ *Thirty-one should appear on every page, because 1 + 2 + 4 + 8 + 16 = 31*

◆ Try to extend the table to produce numbers from 1 to 63. You will need to add an extra page with 32 (2^5) showing in the top left corner.

Mind-reading Page 1

1	23	45
3	25	47
5	27	49
7	29	51
9	31	53
11	33	55
13	35	57
15	37	59
17	39	61
19	41	63
21	43	

Mind-reading Page 2

2	23	46
3	26	47
6	27	50
7	30	51
10	31	54
11	34	55
14	35	58
15	38	59
18	39	62
19	42	63
22	43	

Mind-reading Page 3

4	23	46
5	28	47
6	29	52
7	30	53
12	31	54
13	36	55
14	37	60
15	38	61
20	39	62
21	44	63
22	45	

Mind-reading Page 4

8	27	46
9	28	47
10	29	56
11	30	57
12	31	58
13	40	59
14	41	60
15	42	61
24	43	62
25	44	63
26	45	

Mind-reading Page 5

16	27	54
17	28	55
18	29	56
19	30	57
20	31	58
21	48	59
22	49	60
23	50	61
24	51	62
25	52	63
26	53	

Mind-reading Page 6

32	43	54
33	44	55
34	45	56
35	46	57
36	47	58
37	48	59
38	49	60
39	50	61
40	51	62
41	52	63
42	53	

Number Strips

To perform the following trick you will need to prepare four cardboard strips with the following numbers on the front and back. (See the answers and explanations section for instructions on how to create your own number strips.)

Front

7	3	2	9
6	4	1	8
3	8	7	3
9	4	6	6
1	5	5	2

Back

8	5	3	6
3	0	7	2
4	1	6	8
2	5	9	3
6	9	2	3

Give a student the four number strips and ask him/her to arrange them to form 5, four-digit numbers, e.g.

7	5	6	9
6	8	2	8
3	1	8	3
9	5	3	6
1	9	3	2

Challenge the students to add the five numbers before you do. While they are busy adding the five numbers, simply add 22,220 to the second four-digit number to find the answer.

```
  22,220
+  6,828
  29,048
```

Explanation

This trick works because each strip adds up to 20 plus the value shown in the second row.

7	3	2	9
6	4	1	8
3	8	7	3
9	4	6	6
1	5	5	2
8	5	3	6
3	0	7	2
4	1	6	8
2	5	9	3
6	9	2	3

7 + 3 + 9 + 1 = 20   3 + 8 + 4 + 5 = 20   2 + 7 + 6 + 5 = 20   9 + 3 + 6 + 2 = 20
8 + 4 + 2 + 6 = 20   5 + 1 + 5 + 9 = 20   3 + 6 + 9 + 2 = 20   6 + 8 + 3 + 3 = 20

Regardless of which strips are chosen, they will always add up to 20, therefore adding 22,220 (20,000 + 2,000 + 200 + 20) to the numbers in the second row will give the answer.

You can make your own set of number strips by listing combinations of numbers that add to 20 on each set of eight strips (i.e. 4 front and back), leaving the second row free on each strip. Any number may be written in the second position.

Variations
- The position of the key digits may be altered from the second row to any other row on the strip.
- For younger children it would be appropriate to use 6 strips (i.e. 3 front and back) so that 3 digit numbers are formed. In this case in place of 22,220 you would add 2,220. For example, 3 strips of 4 numbers each and the key row is the third row. The rest of the numbers in the strip add to 20. Why not make up some strips of your own, or use 10 strips to create 5 digit numbers?

Number Strips

7	3	2	9
6	4	1	8
3	8	7	3
9	4	6	6
1	5	5	2
8	5	3	6
3	0	7	2
4	1	6	8
2	5	9	3
6	9	2	3

78 ~ **Magic with Math**   www.didax.com ~ Didax Educational Resources

Blank Strips

Tricky Tables

☞♦ Draw the following table on the chalkboard or a poster.

9	7	5	11
7	5	3	9
13	11	9	15
8	6	4	10

♦ Ask each student to make his/her own copy or photocopy those on the following page.
♦ Give the following instructions:
- Circle any number, then cross out all other numbers in the same column and row as the circled number.

- Circle another number that has not already been circled or crossed out, then cross out all the remaining numbers in the same row or column.

- Repeat the last step.

- You should be left with one number. Circle that number.

- Add all the circled numbers.

♦ Ask your students to try again, but this time to choose different numbers to circle.
♦ What do you notice?
♦ Repeat the steps above using this new table.

13	10	15	12
6	3	8	5
11	8	13	10
4	1	6	3

♦ What do you notice this time?

Explanation

The students should find that their circled numbers always add up to 33. This is because the original table was produced from an addition table. (Note: (5 + 3 + 1 + 7) + (4 + 2 + 8 + 3) = 33)

e.g.

+	5	3	1	7
4	9	7	5	11
2	7	5	3	9
8	13	11	9	15
3	8	6	4	10

To make your own "magic table," simply start with an addition table of your own choice. e.g.

+	4	1	6	3
9	13	10	15	12
2	6	3	8	5
7	11	8	13	10
0	4	1	6	3

Remove the top row and left column to leave a 4x4 table.

You can determine what the "secret total" will be by simply adding the top row and left column of your original table:

(4 + 1 + 6 + 3) + (9 + 2 + 7 + 0) = 32.

Tricky Tables (Cont.)

Variations

- Create a larger table by altering the number of entries in the top row and left column. Note there should always be the same number of rows and columns, so you are left with a square table.
- Create a multiplication table. An example of a multiplication table is shown below. Here the students would have to multiply their final circled numbers.
- Upper elementary teachers might like to create a subtraction table. This would provide good practice in adding negative numbers.

Multiplication table

original table

x	1	3	5	7
2	2	6	10	14
4	4	12	20	28
1	1	3	5	7
6	6	18	30	42

The secret number is:

(1 x 3 x 5 x 7) x (2 x 4 x 1 x 6) = 5,040.

The steps are the same as in the previous example, except that when all the circled numbers have been found, they are multiplied instead of being added.

2	6	10	14
4	12	20	28
1	3	5	7
6	18	30	42

6 x 3 x 10 x 28 = 5,040

Tables

9	7	5	11
7	5	3	9
13	11	9	15
8	6	4	10

13	10	15	12
6	3	8	5
11	8	13	10
4	1	6	3

9	7	5	11
7	5	3	9
13	11	9	15
8	6	4	10

13	10	15	12
6	3	8	5
11	8	13	10
4	1	6	3

9	7	5	11
7	5	3	9
13	11	9	15
8	6	4	10

13	10	15	12
6	3	8	5
11	8	13	10
4	1	6	3

Lightning Addition

	Example
	2, 6

◆ Ask each student to write down any two numbers.
(The first time you try this trick it is a good idea to start with two small numbers.)

◆ The idea is to create a sequence of ten numbers using the chosen numbers as the starting point.

◆ To create the third number in the sequence, add the first two numbers. $2 + 6 = 8$

◆ Now add the second and third numbers to create the fourth. $6 + 8 = 14$

◆ Continue this way until you reach the tenth number in the sequence.

2, 6, 8, 14, 22, 36, (58), 94, 152, 246
   seventh number

◆ Ask each student to show you the set of numbers he/she created for a short time (a few seconds is enough). Note the seventh number in the sequence (in this case 58).

◆ Instruct the student to add his/her ten numbers.

◆ While the students are performing the addition, mentally multiply the seventh number in the sequence by 11 and casually write the number on the board.

Explanation

You may recognize this sequence as a Fibonacci sequence.
If we start with the sequence:

1 a
2 b
3 $a + b$
4 $a + 2b$ or $(a + b) + b$
5 $2a + 3b$ or $(a + 2b) + (a + b)$
6 $3a + 5b$ or $(2a + 3b) + (a + 2b)$
7 $5a + 8b$ or $(3a + 5b) + (2a + 3b)$
8 $8a + 13b$ or $(5a + 8b) + (3a + 5b)$
9 $13a + 21b$ or $(8a + 13b) + (5a + 8b)$
10 $21a + 34b$ or $(13a + 21b) + (8a + 13b)$

11 x (seventh number) = total
11 x $(5a + 8b)$ = $55a + 88b$

To multiply a two-digit number by 11:
Take the two digits of the original number to form the first and last digits of the answer. The middle digit(s) is found by adding the two digits:
e.g. 36 x 11
     3 _ _ 9 (3 + 6 = 9)
gives 396
When the two digits add up to more than 9 carry a one to the next number:
e.g. 89 x 11
     8 _ _ 9 (8 + 9 = 17)
     979

A Fibonacci sequence of numbers is one in which the third number is the sum of the first two, the fourth is the sum of the second and third number and so on. Put simply, a term in the sequence is found by adding the two previous terms. While the sequence shown in "Lightning Addition" is a Fibonacci sequence it is not the Fibonacci sequence which begins with 1, 1, 2, 3, 5, 8, 13, 21, 34, 55, 89...

Super Sequences

- ◆ *Ask a student to write down any number; e.g. 10.*
  (The first time you try this trick it is a good idea to start with a number less than 20. If larger numbers are allowed you will probably need a calculator.)
- ◆ Next, ask the class to add all the numbers up to, and including, the chosen number.
  $1 + 2 + 3 + 4 + 5 + 6 + 7 + 8 + 9 + 10 = 55$
- ◆ While the class is hard at work adding the numbers, casually write the answer on the chalkboard.
- ◆ To work out the answer quickly, multiply the last number in the sequence by one more than itself, then halve the result.
  $10 \times 11 = 110$ then $110 \div 2 = 55$

Explanation

The rule for working out the sum of the numbers in a sequence starting at one and increasing by one each time is:

$$\frac{n(n+1)}{2}$$

That is, multiply the last number in the sequence by one more than the number, then halve the result.

After trying this trick a few times, you may wish to inject some history into the lesson by discussing the famous mathematician named Gauss. Gauss was a bright, but very troublesome, student. One day, his teacher decided to keep young Gauss quiet for a while. He gave him the task of adding all the whole numbers from 1 to 100. There were no calculators back in that time, so the teacher expected Gauss to add the numbers manually. Imagine the surprise on the teacher's face when, in less than a minute, Gauss returned with the answer. Here is how he did it. He noticed the following pattern:

$1 + 2 + 3 + ... + 98 + 99 + 100$

Each pair of numbers added to 101. There were 50 pairs, therefore the total was 5,050!

Variations

- Try adding all the odd numbers up to a particular number; e.g. $1 + 3 + 5 + 7 + 9 + 11...$
- To find the sum of consecutive odd numbers in a sequence, add 1 to the largest number, halve it and then square it.
- Try using a sequence of consecutive even numbers; e.g. $2 + 4 + 6 + 8 + 10$.
- To quickly find the sum of consecutive even numbers, halve the largest number and multiply the result by one more than itself.

Mystifying Multiplication

- ♦ Ask one of your students to write a two-digit, three-digit, or larger multiplication problem on the board.

    4,362 (multiplicand)
    x 7,746 (multiplier)

- ♦ Now write your own calculation, which contains the same multiplicand on the board.

    4,362 (multiplicand)
    x 2,253 (multiplier)

- ♦ The multiplier you choose should be the difference between 9,999 and the original multiplier (i.e. 9,999 – 7,746 = 2,253).
- ♦ Announce that you will perform both multiplications and add them together in your head while the rest of the class does the same on a calculator.
- ♦ The answer may be found by subtracting 1 from your multiplicand (e.g. 4,362) to produce the first digits of your answer (4,361). The last four digits may be found by subtracting this number (4,361) from 9,999 (9,999 – 4,361 = 5,638). The answer to the problem shown above would be 43,615,638.

Explanation

To make this trick work, you must subtract the student's multiplier from 9,999:

e.g.
```
  9,999
- 7,746
  -----
  2,253
```

and use the result as your multiplier. In other words, the sum of the two multiplications will be the same as multiplying the original multiplicand by 9,999. All you have to do to arrive at the answer is to subtract 1 from your multiplicand to find the first set of digits in your answer and then subtract this number from 9,999 to find the last four digits in the answer.

The answer for 4,362 x 7,746 and 4,362 x 2,253 is 43,615,638.

Mathematically, $n \times 9{,}999 = (n \times 10{,}000 - n)$
$= (n - 1) \times 10{,}000 + [9{,}999 - (n - 1)]$.

Variation: the same principle works on two- and three-digit numbers.

Number Spelling

- Write the following set of numbers on the chalkboard or a poster. The order is not important.

2	18	4
21	15	24
8		12

- Ask one of the students to choose a number while you turn your back.
- Ask the rest of the class to spell out the number in their heads while you point to various numbers on the board.
- They should spell a letter to themselves every time you point to a number.
- The order in which you point to the numbers is important. (See the answers and explanations for the technique to use.)
- Instruct the class to call out stop when the finish spelling the word. You should find that you are pointing at the secret number.

Explanation

The set of numbers chosen is very special. There is a relationship between the number of letters used to spell each number.

TWO (3)   FOUR (4)   EIGHT (5)   TWELVE (6)   FIFTEEN (7)
EIGHTEEN (8)   TWENTY ONE (9)   TWENTY-FOUR (10)

To make the trick work, point to any two numbers at random and then point to the others in the order given above. You will note that in this case, the number of letters and the actual numbers are in order. The third number you should point at is 2, then 4, 8 and so on. If the number chosen was 15 then the seven numbers will have been pointed to by the time the students have spelled out the seven letters of the word fifteen and you will be pointing at 15.

You may like to vary the numbers.

2	17	11
7	13	
16	25	5

10	19	8
22	16	29
4	20	

The arrangement is not important but the pointing order is.

It all Adds up

- ☛ ♦ Ask your students to choose a four-digit number. e.g. 3,712
- ♦ Write it on the board.
- ♦ Ask the students to choose another four-digit number and write it on the board. 4,653
- ♦ Choose a four-digit number yourself and write it on the board. 5,346
  (Choose your four-digit number so that with their second four-digit number it totals to 9,999).
- ♦ Ask the students to choose another four-digit number. 6,719
- ♦ Choose another four-digit number yourself. 3,280
  (Again, choose a four-digit number so that with their third number it adds to 9,999.)
- ♦ Ask the students to add the five numbers.
- ♦ While they are working out the sum, simply place 2 in front of the original number (e.g. 23,712) and then take 2 away from the one's digit (e.g. 23,710). This will give the answer much more quickly than the students can add the numbers.
- ♦ Casually write it on the board.
- ♦ This trick may be easily altered to work with three- and five-digit numbers.

Notes

About the Author, Paul Swan

Paul Swan has worked as an elementary and high school teacher and now works as a teacher educator at Edith Cowan University, Perth, Western Australia. He was awarded his Ph.D. for his research in children's computation choices and methods.

He has a passion for mathematics and is always looking for ways to make mathematics interesting. He writes mathematics books in his spare time, as well as presenting conference workshops, teacher professional development seminars and mathemagic shows across Australia.

As a father of four boys – all in school – he recognizes the challenges faced by teachers trying to motivate children to learn mathematics. As a result, he tries to make his publications stimulating and inviting for children, while at the same time challenging them to think.